GEOGRAPHY AND TRANSITION IN THE POST-SOVIET REPUBLICS

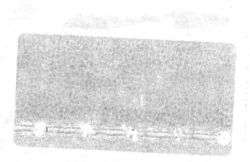

GEOGRAPHY AND TRANSITION IN THE POST-SOVIET REPUBLICS

Edited by Michael J. Bradshaw

School of Geography
University of Birmingham, UK

JOHN WILEY & SONS
Chichester • New York • Weinheim • Brisbane • Singapore • Toronto

Published 1997 by   John Wiley & Sons Ltd,
Baffins Lane, Chichester,
West Sussex PO19 1UD, England

National   01243 779777
International (+44) 1243 779777
e-mail (for orders and customer service enquiries): cs-books@wiley.co.uk
Visit our Home Page on http://www.wiley.co.uk
or http://www.wiley.com

Other Wiley Editorial Offices

John Wiley & Sons, Inc., 605 Third Avenue,
New York, NY 10158–0012, USA

VCH Verlagsgesellschaft mbH, Pappelallee 3,
D-69469 Weinheim, Germany

Jacaranda Wiley Ltd, 33 Park Road, Milton,
Queensland 4064, Australia

John Wiley & Sons (Asia) Pte Ltd, 2 Clementi Loop #02–01,
Jin Xing Distripark, Singapore 129809

John Wiley & Sons (Canada) Ltd, 22 Worcester Road,
Rexdale, Ontario M9W 1L1, Canada

Library of Congress Cataloging-in-Publication Data

A catalogue record for this book is available from the Library of Congress

British Library Cataloguing in Publication Data

A catalogue record for this book is available from the British Library

ISBN 0-471-94891-8 (cloth)
ISBN 0-471-94892-6 (paper)

Typeset by Dorwyn Ltd, Rowlands Castle, Hampshire
Printed and bound in Great Britain by Bookcraft (Bath) Ltd
This book is printed on acid-free paper responsibly manufactured from sustainable forestation,
for which at least two trees are planted for each one used for paper production.

Contents

Preface

Had the World been a different place, this book should logically have been the second edition of *Soviet Union: a new regional geography?* That book was published in the autumn of 1991, just as the Soviet Union was in its final death throes. The current collection of essays is the successor to the first book in that it seeks present research on the relationship between geography and societal change in the, now, *former*-Soviet Union. The context for that research has changed dramatically, the Soviet Union has been replaced by 15 independent republics, each of which is pursuing its own independent path towards some form of market-type economic system.

Now, perhaps more than ever before, geography matters as the progress of each of the post-Soviet Republics, and their constituent regions, is shaped by the legacies and endowments of the past as well as their relative location and potential in the global system. Five years on, the post-Soviet Republics are clearly lagging behind the economies of East/Central Europe. At the same time, there is growing differential between the republics, the Baltic States have been relatively successful at distancing themselves from the rest of the former Soviet Union; while the Transcaucasus has suffered from internal conflict. Similar processes of differentiation are found within the larger republics, such as the Russian Federation, Ukraine and Kazakhstan. As the scale of control has devolved from Moscow, to republican governments and also to the regions, so the level of economic activity and the socio-economic conditions within individual republics and regions has been determined by the resources at the immediate disposal of the political leadership. Thus, inherited economic structure, resource endowment, political status and access to external sources of supply and markets have all become key factors in determining the relative status of the post-Soviet Republics.

As a consequence of the collapse of the Soviet Union, interest in the geography of the post-Soviet Republics has increased significantly. Unfortunately, this has happened at a time when the sub-discipline of "post-Soviet geography" is suffering a flight of human capital. Many scholars in the West previously active in the field have taken opportunities in the private sector and in government; while in Russia, for example, the financial situation has forced many to leave academia to find alternative ways of making a living. While a new generation of specialists is active at the postgraduate level, there is no guarantee that they will find the security of a university post from which they can develop their research careers. Thus, it is frustrating that we find ourselves in a situation where there are many exciting research issues to pursue, but there are now fewer established scholars to pursue them.

Like the previous book, this publication has its origins in a series of special sessions at the Association of American Geographers Annual Meetings, this time in San Francisco in the spring of 1994. That this book took longer to come to fruition is in part due to the logistical problems of bringing together contributors from far afield, but also because the list of contributors was constantly changing. Nevertheless, the book succeeds in conveying the key dimensions of the relationship between geography and transition since 1991. The emphasis is very much upon the politics of state formation and national identity. At the same time, the importance of the legacy of the Soviet era emerges as a constant theme, as does the process of the decentralisation of decision making. A conscious attempt has been made to extend the geographical coverage beyond Russia, but it remains the case that the majority of post-Soviet geographers in the West are involved in research on Russia.

Reviews of the first book suggested that the publisher's description of it as a text book caused problems. Those teaching courses on the, then, Soviet Union seemed unsure how to use the book. The intent was, and remains, to provide an interface between the teaching and research. It is assumed that the readership is familiar with the basic geography of the post-Soviet Republics (key texts for such a course would be: Bater 1996 and Shaw 1995). The chapters in this book present research on key aspects of transition, they also provide the reader with access to the specialist literature. It is also hoped that the book will find a readership in systematic courses in geography, especially those in political geography concerned with nationalism and identity and economic geography concerned with economic restructuring. Finally, it is hoped that the book can provide area specialists in other disciplines with a starting point for considering the importance of geographical processes in the transformation of the post-Soviet Republics.

The individual contributors have acknowledged the support of various institutions, as editor, I would like to thank the contributors for meeting the various deadlines and responding to my queries so promptly. Any errors that remain are the responsibility of the editor. I would also like to thank Kevin Burkhill and David Haddock in the School of Geography for their assistance in producing some of the maps used in this book. The first book came out weeks after the aborted August coup in 1995, even as I write there are doubts over the future of President Yeltsin and the Russian Army is feared to be on the verge of revolt. Thus, the transition remains far from complete and this book is far from the last word on the changing geography of the post-Soviet Republics.

Michael J. Bradshaw
November, 1996
School of Geography
The University of Birmingham, UK

Notes on Contributors

Michael J. Bradshaw is a Senior Lecturer in the School of Geography and an Associate Member of the Centre for Russian and East European Studies at the University of Birmingham, UK. He is editor of *The Soviet Union: a new regional geography?* and author of *Russia's regions: a business analysis*.

Ralph Clem is Professor of Geography at Florida International University in Miami, Florida, USA. He is the co-author and editor of books and articles dealing mainly with demography and ethnicity in the former Soviet Union. Most recently he has co-authored a series of articles in *Post-Soviet Geography* examining the electoral geography of post-Soviet Russia.

Peter De Souza is Associate Professor in the Department of Economic Geography, Gothenborg University, Sweden. His research interests are in the Russian Far East, military conversion and centre–periphery relations in Russia. He is the author of *Territorial–Production Complexes in the Soviet Union – with special focus on Siberia*.

Robert J. Kaiser is Assistant Professor in the Department of Geography at the University of Wisconsin-Madison, Wisconsin, USA. He is author of *Geography of nationalism in Russia and the USSR* and co-author of *The Russians as a new minority in the newly independent states of the former Soviet Union*.

Nicholas Lynn is Lecturer in the Department of Geography at the University of Edinburgh, UK. He recently received a PhD from the University of Birmingham for a thesis entitled: *A Political Geography of the Republics of the Russian Federation*.

Robert N. North is Associate Professor of Geography in the Department of Geography at the University of British Columbia, Vancouver, British Columbia, Canada. He is presently working on a book on the role of transport in the development of the Russian North and is author of *Russian Transport: problems and prospects*.

Beth Mitchneck is Assistant Professor in the Department of Geography and Regional Development at the University of Arizona, Tucson, Arizona, USA. Her research interests are regional development, interregional migration and local government. She has published articles on these themes in such journals as *Economic Geography* and *Professional Geographer*.

Alexei Novikov is Lecturer in the Faculty of Geography at Moscow State University, Russia. He has published numerous articles on aspects of Federalism and cultural geography. He is editor of the journal *Geograffiti*.

Judith Pallot is a Lecturer in the School of Geography and an Official Student of Christ Church at the University of Oxford, UK. She is presently completing a book on the Stolypin reforms and conducting research on rural change in Russia. She is co-author of *Landscape and Settlement in Romanov Russia*.

Philip R. Pryde is a Professor in the Department of Geography at San Diego State University, California, USA. He is author of *Environmental Management in the Soviet Union* and editor of *Environmental Resources and Constraints in the former Soviet Republics*.

Denis J.B. Shaw is Reader in Russian Geography in the School of Geography and an Associate

Member of the Center for Russian and East European Studies at the University of Birmingham, UK. He is editor of *The Post-Soviet Republics: a systematic geography*.

Graham Smith is a Fellow of Sidney Sussex College and a Lecturer in the Department of Geography at the University of Cambridge, UK. He is editor of *The nationalities question in the post-*

Soviet states and *The Baltic States: the national self-determination of Estonia, Latvia and Lithuania*.

Alison C. Stenning is a PhD Candidate in the School of Geography at the University of Birmingham, UK. She is the recipient of an Economic and Social Research Council Research Studentship to examine local economic development in Novosibirsk, West Siberia, Russia.

1

Introduction: transition and geographical change

Michael J. Bradshaw
University of Birmingham, UK

The prevailing assumption in the social sciences is that society and economy have geographical outcomes but not geographical foundations. We disagree. In our view the territorial arrangement of activities is central to the broader constitution of any society's economic, social, and political fabric; indeed, societies are shaped only by virtue of their imbrication in territorial formations (Storper and Walker, 1989, p. 226).

Open any recent book on the former Soviet Union and the chances are that there will be at least one map. This concern with geography is a relatively recent phenomenon within what used to be called Soviet studies. A few years ago each book contained the obligatory diagram of the structure of the Communist Party of the Soviet Union (CPSU). The fact that one now needs a map to navigate events in post-Soviet studies should be evidence enough that geography matters! The aim of this collection of essays is to explore the relationship between geography and transition.

Much of current analysis in post-Soviet studies uses terms such as "region" and "locality" as if they are unproblematic. As human geographers are well aware, this is far from the case. The contemporary geographical literature is rich with discussions about the meaning of place and the nature of locality, about the relationship between the global and the local and the cultural and the economic. The processes of systemic transformation now under way in the post-Soviet republics are forcing area specialists to become familiar with new dimensions of the social sciences. If it is necessary to turn to economics to know more about privatisation, surely social scientists interested in regional dimensions of transition should turn to contemporary human geography to understand issues related to territory and identity or regional change. In other words, I am suggesting that in order to understand the regional dimensions of systemic transformation in the former Soviet Union it is necessary to know something about the contemporary human geography of the post-Soviet republics (see Shaw, 1995 for a systematic review).

It may well come as a surprise to former Soviet studies (or whatever it is now called) that there is far more to contemporary human geography than a catalogue of facts about what happens where (trivial pursuits geography). Today human geography is concerned with the spatiality of social processes. As Storper and Walker (1989, p. 226) suggest, geography matters because it is a crucial component of societal change. Nowhere is this more evident than in the post-Soviet states of East/Central Europe and the former Soviet Union. Geographical variations in resource endowment, the complex mosaic of nationalities, the environmental legacies of rapid industrialisation, the distribution of defence production, and the importance of regional identity in the cultural and political construction of nationality, all of these layers, and many more, shape attitudes to, and the progress of, systemic transformation. This introductory chapter serves to flag the key issues that are examined in the essays that follow.

Geography and transition

In the concluding chapter of *The Soviet Union: A New Regional Geography?*, in fact in the last sentence of the book, I wrote: "the challenge now is to come to terms with the process of transition" (Bradshaw, 1991, p. 194). Transition can be defined as "change from one state to another". At the time of writing I was, of course, referring to a transition from a planned economy to a market-oriented economy. At that time I had no idea that transition in the Soviet context would come to mean change from one state to 15!

The attention of Western academics and policy makers has tended to be focused upon the progress of economic transition in the former Soviet Union, but the reality is that economic transition is but one component of a more complex process which is often termed *systemic transformation*. The process of systemic transformation encompasses at least three fundamental challenges to the post-Soviet republics: a *political transition* that involves the creation of independent states, the establishment of state identity and the formation of foreign policy; an *economic transition* that involves the liberalisation of prices, the

Geography and Transition in the Post-Soviet Republics. Edited by M.J. Bradshaw.
© 1997 M.J. Bradshaw & Contributors. Published 1997 by John Wiley & Sons Ltd.

stabilisation of government expenditure, the privatisation of state-owned enterprises and the internationalisation of the economy; and a *social transition* that involves the establishment of democratic institutions and the development of civil society (Bradshaw, 1993, p. 1). Yergin and Gustafson (1994), with reference to Russia, talk of "the triple transition": from dictatorship to democracy; from command economy to free market; and from empire to nation-state. Of course, these various aspects of transition are intimately entwined and at times exert contradictory influences upon the process of systemic transformation. For example, the need for cooperative action to manage inherited dependencies, and thus smooth economic transition, is undermined by the state-centred policies pursued by the individual new republics to establish an independent political identity. Similarly, the overthrow of democratic institutions has been justified by the need to protect the progress of economic reform. At the same time, the dynamics of the dimensions of transition differ, both within a particular state and also between the states. In Russia, the progress of economic reform has been hampered by political stalemate between parliament and the president, a problem resolved by the shelling of the White House, by bargaining between the centre and the regions over the implementation of the new constitution, and most recently by the successes of the new Russian Communist Party. However, compared to the progress of economic reform in Ukraine, Russia is moving ahead. The European Bank for Reconstruction and Development's (EBRD, 1995) latest *Transition Report* reveals that while Russia is ahead of the rest of the Commonwealth of Independent States in terms of progress in economic transition, the fast-reforming economies of Central Europe are racing ahead of Russia. The Baltic states occupy an intermediary position between Central Europe and Russia, while the republics of the Transcaucasus and Central Asia are at the bottom of the table. Variations in the pace of transformation are to be expected as the post-Soviet republics did not all start from the same point (Bradshaw and Lynn, 1994). The Baltic states, for example, were the most "developed" Soviet republics and have benefited from the fact that they have previously experienced an existence as independent sovereign states; elsewhere, in Central Asia for example, state formation is a new experience. The very different circumstances of the individual states should guard against the imposition of a general model of transition for the prescription of acceptable outcomes. Despite what Western politicians may desire, the collapse of the Soviet Union will not produce 15 Western-style market-oriented democracies.

Before considering the relationship between economic transition and geographical change, it is necessary to spend some time examining the notion of transition. The idea of moving from one state to another suggests an orderly progression from one steady state, in this case a planned economy, to a new steady state, in this case a market economy (see Figure 1.1). In such a scheme the process of transition is the means by which the planned economy is transformed into a market economy. In the planned economy regional development was the consequence of highly centralised economic development along sectoral lines; in marked contrast, the processes of systemic transformation are reducing the level of central control and empowering regional authorities and individual enterprises, as well as individual citizens who now have to make decisions previously made for them by the state. Thus the scale at which decisions are made has been devolved to the local and individual scale.

In the past for the Soviet geographer the task was to describe the geography produced by the

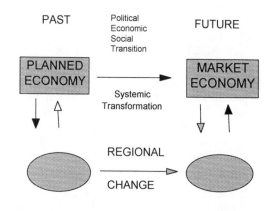

Figure 1.1 *Transition and regional change*

planned economy. At its best, Soviet geography explored the ways in which the planned economy produced a particular pattern of settlement, distribution of economic activities, pattern of regional development, transportation system, and so on (see e.g. Dienes, 1987; Pallot and Shaw, 1981). That knowledge is far from redundant as it is the landscape shaped by the planned economy that is influencing and constraining the process of systemic transformation today. At the same time, as we find out more about the nature of the Soviet Union, a new geography is revealed which requires that we reassess the Soviet experience and the production of a Soviet geography.

There may be those who feel that we must wait until a new "steady state" is created in the post-Soviet republics before we can again describe their geographies. Like the "before" and "after" pictures depicting a dramatic change in waistline, this kind of geography has nothing to say about the dynamic processes that take us from one state to the next. I reject such a view on two grounds: first, because it relegates the post-Soviet geographer to a purely descriptive role, which places him or her at odds with contemporary human geography; and second, because geography is itself a crucial component of the process of transition. Transition is spatially constituted in the sense that geography constrains what is possible. The spatial dimensions of transition influence and are influenced by the existing landscape; geography offers constraints and opportunities, a spatial structure within which human agency operates. Transition from a planned economy to a market economy involves a fundamental restructuring of the spatial division of labour within the post-Soviet republics and simultaneously, a reorientation of the territorial division of labour between those states. At the same time, the incorporation of 15 new states into the global political–economic system has implications for the entire system: new markets are being developed and new sources of competition are emerging. Thus, the collapse of the Soviet Union has implications for global order and anyone interested in global geographies cannot afford to ignore the dramatic changes now taking place in the post-Soviet republics.

Geographies of transformation

The complexities of systemic transformation defy the imposition of any single approach to their analysis. Thus, the essays presented in this book can be seen as presenting different approaches to various geographies of transformation. This is in part due to that fact that the authors themselves span a number of generations of geographers, from those who were steeped in the Soviet system to those whose research careers are post-Soviet. It is also due to the fact that different dimensions of transformation naturally lead to different approaches to examining geographical change, some contributions are more theoretical, while others focus on the empirical detail of change in the post-Soviet republics. In no sense can this book claim to have covered all possible geographies of transformation. No doubt there are a variety of different approaches that could be applied to the issues raised in the various chapters; equally there are many aspects of geographical change in the post-Soviet republics that are not addressed by this book. As with the previous collection of essays, *The Soviet Union: A New Regional Geography?*, the aim is to present a collection of essays by geographers who are seeking to develop new ways of thinking about the former Soviet Union. Therefore, to a large degree, the scope of the book has been determined by the research interests of those working on the region. Comparison with the previous book will reveal that some of the original authors are no longer involved in this project to construct a new regional geography of the, now, post-Soviet republics. In every instance this is because their careers have taken them in different directions, into university administration, government organisations or the private sector. Faced with this problem of human capital flight, the current collection of essays is more international, including authors from North America, Western Europe and Russia. This multinational aspect to the book introduces a further dimension to the various geographies of transformation (as well as problems of translation).

Despite the variety of issues, approaches and authors, the book does focus upon many of the key dimensions of transformation. The first half of the book focuses upon issues of nationality, territoriality and identity. In Chapter 2 Robert Kaiser re-evaluates the issues of national identity and nationalism in the Soviet Union and examines sources of tension and conflict in the post-Soviet republics (see also Kaiser, 1994). A review of recent developments leads to the development of a typology of nationalisms in the newly independent states. Chapters by Shaw and Novikov present different dimensions of national identity in the Russian Federation. In Chapter 3 Denis Shaw adopts a long-term historical approach to understanding the identity crisis facing Russia. In Chapter 4 Alexei Novikov also addresses the issue of Russian national identity, but he presents a cultural argument for Russians' apparent lack of a sense of place. Chapter 5, by Nicholas Lynn, focuses upon the current territorial–administrative structure of the Russian Federation and the special place occupied by Russia's republics. Analysis of Russia's republics reveals considerable variation in the level of economic development, resource endowment and nationalism which translates into a variety of different relations with the federal government in Moscow. In Chapter 6 Graham Smith considers the plight of the Russian diaspora in the "near-abroad". It is argued that there is a need to recognise the variety of positions of ethnic Russians in the various post-Soviet republics. The chapter then focuses upon the situation in the Baltic states of Latvia and Estonia and the peculiar situation in Narva in Estonia, a region with a Russian majority (see also Smith, 1994, 1996). Chapter 6 can also be read as an alternative approach to the issues of nationalism raised by Robert Kaiser in Chapter 2. All of these essays stress the geographical complexity of the situation in the post-Soviet republics and the dangers of overgeneralisation.

The second half of the book focuses upon a variety of systematic aspects of transformation. Chapters 7 and 8 examine the impact of systemic transformation in Russia's cities and in the Russian countryside. Both chapters consider the impact of decentralisation. Beth Mitchneck focuses on the transfer of responsibility to local governments, while Judy Pallot examines various attempts to break up collectivised agriculture. The dramatic changes taking place in city government contrast with the stubborn inertia of those in the countryside. This ideological divide is clear in Russia's political geography, the major cities being bastions of reform, while the countryside remains resolutely communist. Local government decentralisation is seen as a return to a Tsarist past and a rejection of the Soviet system, in the countryside resistance to decentralisation can be interpreted as a desire to retain the Soviet system. Chapters 10 and 11 consider the legacies of Soviet industrialisation. In Chapter 10 Philip Pryde surveys the environmental problems facing the post-Soviet republics (see also Pryde, 1995). From this analysis, it is clear that those republics with the greatest problems are those least able to afford the cost of clean-up. Equally, it is clear that many of the environmental legacies of the Soviet Union are now problems that are international in scale. In Chapter 10 Alison Stenning considers the relationship between economic restructuring and regional change in Russia. The chapter makes a theoretical argument concerning the nature of economic transition and the relationship between the various elements of economic transition and regional economic performance. Chapters 11 and 12 present regional case studies. In Chapter 11 Ralph Clem surveys the prospects for development in post-Soviet Central Asia, while in Chapter 12 Peter de Souza considers the new geoeconomic position of the Russian Far East. In both cases the collapse of the Soviet Union has resulted in a process of reconfiguration. In Central Asia the various republics are having to come to terms with independent identities and are having to shape their own development strategies to address the substantial problems that they all face. At the same time, these republics are having to forge new relations with each other and with their neighbours. Similarly, the Russian Far East finds itself distanced from the economic core of European Russia and is seeking to develop new relations with its immediate neighbours in the Asia–Pacific region. Peter de Souza employs the metaphor of a gateway to examine the various issues that influence the Far East's ability to benefit from its geographical juxtaposition to one of the most important arenas of economic growth in the global economy. The final chapter by Robert North presents evidence for the scale of geographical reorientation that is taking place in the post-Soviet republics. Chapter 13 details how the disintegration of Soviet economic space, economic crisis, privatisation and restructuring have all impacted on the demand for transportation services. Equally, the legacies of the Soviet system mean that the existing transportation system is unable to meet the demands of the emerging market economy. In a world of "time-space compression" it seems that in Russia and the other post-Soviet republics the friction of distance is actually increasing. Thus the political and ethnic fragmentation of the post-Soviet republics is paralleled by the decentralisation and regionalisation of economic activity. This fragmentation of the Soviet system is the reason for the heightened interest in the geographical aspects of systemic transformation.

References

Bradshaw, M.J. (ed.) (1991), *The Soviet Union: A New Regional Geography?* London: Belhaven Press.

Bradshaw, M.J. (1993), *The Economic Effects of Soviet Dissolution.* London: RIIA.

Bradshaw, M.J. and Lynn, N.J. (1994), "After the Soviet Union: the post-Soviet states in the world system", *The Professional Geographer*, **46**(4): 439–49.

Dienes, L. (1987), *Soviet Asia: Economic Development and National Policy Choices.* Boulder, Colo.: Westview Press.

EBRD (1995), *Transition Report 1995: Investment and Enterprise Development.* London: EBRD.

Kaiser, R. (1994), *The Geography of Nationalism in Russia and the USSR.* Princeton: Princeton University Press.

Pallot, J. and Shaw, D.J.B. (1981), *Planning in the Soviet Union.* London: Croom Helm.

Pryde, P.R. (1995), *Environmental Resources and Constraints in the Former Soviet Republics.* Boulder, Colo.: Westview Press.

Shaw, D.J.B. (1995), *The Post-Soviet Republics: A Systematic Geography.* Harlow: Longman.

Smith, G. (ed.) (1994), *The Baltic States: The National Self-determination of Estonia, Latvia and Lithuania.* London: Macmillan.

Smith, G. (ed.) (1996), *The Nationalities Question in the Post-Soviet States.* Harlow: Longman.

Storper, M. and Walker, R. (1989), *The Capitalist Imperative: Territory, Technology, and Industrial Growth.* Oxford: Blackwell.

Yergin, D. and Gustafson, T. (1994), *Russia 2010: and What it Means for the World.* New York: Random House.

2
Nationalism and identity

Robert J. Kaiser
University of Wisconsin-Madison, USA

National identity and nationalism have become the cornerstones on which sociocultural, economic and political relations in the newly independent states (NIS) of the former USSR are being built. Yet it was not so long ago that nations in the USSR were treated as communities facing extinction, and the consensus among Western and Soviet scholars was that nationalism itself was non-existent, or was fast becoming an irrelevant ideology in the USSR. Studies of the so-called national question in the USSR, particularly those conducted prior to the 1980s, clearly misread societal trends in reaching these conclusions. This chapter re-examines the national question in light of what has occurred since the late 1980s, and offers an alternative interpretation of the changing nature of national identity and nationalism in the post-Second World War Soviet Union. On the basis of this alternative interpretation, the chapter will examine the rise in international tensions and conflicts that have accompanied the disintegration of the USSR into 15 national states, and assess the prospects for conflict management in the NIS.

Nations and nationalism defined

Before beginning this analysis, it is necessary to define nation and nationalism, since these terms have been used as synonyms for a variety of identities and ideologies. The varying usage of these terms has added to the confusion surrounding national identity and nationalism, and undoubtedly has contributed to the misinterpretation of the impact of societal trends on national identity (Connor, 1978).

National identity

In this chapter, the term "nation" is defined as a mass-based community of belonging and interest, whose members share a backward-looking sense of common genealogical and geographic roots, as well as a forward-looking sense of destiny. As a community of belonging, members

typically view the nation as an extended family related by common ancestry, although this belief in a common ancestor is based more on myths and legends than on an objective appraisal of the nation's history (Connor, 1978; Smith, 1986). Most nations are products of interethnic integration. The myth of common ancestry is critically important, however, because it lends to the nation the appearance of an ascriptive identity (i.e. that one is born with a national identity), and reduces the likelihood that nations can be unmade. It makes nations appear as primordial communities that are both natural and eternal. This primordialist depiction of national identity is emphasised by nationalists.

National origin myths also typically stress the importance of the nation's geographic roots in some ancestral homeland, and often depict the nation as a product of both blood and soil (Williams and Smith, 1983; Connor, 1986; Smith, 1986; Anderson, 1988; Kaiser, 1994a). This myth of primordial connectedness with the homeland serves as one of the main bases for nationalistic claims to territory today. For example, Ukrainian nationalists proclaimed the exclusive right of the Ukrainian nation "to dominion over territory it has settled from time immemorial" (RUKH, 1989, p. 15). Throughout the territory of the former USSR, nationalist historians, archaeologists, anthropologists, sociologists and geographers are searching for evidence that reinforces the nations' primordial claims to their ancestral homelands.

The nation is more than a backward-looking community of belonging; it is also a forward-looking community of interest. This future orientation transforms the nation from a backward-looking ethnocultural community concerned with preserving the past into a politicised interest group intent on seizing control of its fate or destiny (i.e. national self-determination).

This forward-looking aspect of national identity also has a geographic dimension, since most nationalists assert that in order for the nation to gain control over its destiny, it must gain control over some geographic place. Territory becomes the means through which the nation will fulfil its destiny. Of course, the geographic place that

Geography and Transition in the Post-Soviet Republics. Edited by M.J. Bradshaw.
© 1997 M.J. Bradshaw & Contributors. Published 1997 by John Wiley & Sons Ltd.

nationalists assert control over is the ancestral homeland. In this way, the backward- and forward-looking dimensions of national identity are intimately connected through the soil of the homeland. Controlling the ancestral homeland has become the critical political action programme for nationalists during the twentieth century. Without such control, or so nationalists argue, the nation is without a future, since only nations in control of their homelands can ensure their future viability. With such political geographic control over the homeland, again according to nationalists, the nation's future will not only be secured, it will be glorious. This perception was a nearly ubiquitous feature of the national independence movements in the former USSR during the late 1980s and early 1990s. Nationalists blamed all socioeconomic problems on exploitation at the hands of Moscow and the Russians, and talked in glowing terms about the future paradise that would be created once the nation gained control over its own fate in its own homeland.

The nation is not an ethnic group, nor is it a state, although it is often used as a synonym for one or the other. However, as can be seen in the definition and discussion provided above, the nation is intimately related to both ethnic groups and states. Nations may be seen as ethnic groups that have become forward-looking politicised and territorialised interest groups (e.g. Smith, 1986). In turn, as a politicised interest group whose members strive for self-determination in their own homeland, the nation is a community seeking a state of its own (i.e. a nation-state) (Breuilly, 1993).

The nation is also often referred to as a cultural community whose members share a set of tangible traits or objective characteristics, such as language, religion, customs and so forth. Although this is normally the case, the retention of these objective characteristics is not a necessary condition for the maintenance of national identity, and the existence of a community whose members share a common language, religion, etc. is not a sufficient condition in and of itself for the emergence of a national identity. Indeed, the loss of one's native language or religious affiliation, particularly if that loss is part of a concerted effort at forcible assimilation, has often led not to the demise of one's national identity, but conversely to a rise in national self-consciousness as a reaction against the forcible nature of such assimilation processes (Connor, 1972; Kaiser, 1994a).

Nationalism

The term "nationalism" is used in this chapter to refer to both an ideology and a political action programme, each of which is centred around the nation as defined above. As an ideology, nationalism represents an assertion not only that the nation exists, but also that it is the community that matters most to its members (Breuilly, 1993, p. 2). Nationalism demands unconditional loyalty and potentially unlimited sacrifice from the national membership. In return, nationalism promises to members of the nation a glorious future, a heaven on earth. In this way at least, nationalism may be seen as a new religion, which like communism rejects a fatalistic outlook in favour of a more proactive attempt to determine or control fate.

As a political action programme, nationalism represents an effort to mobilise members of the nation behind the future-oriented goal of national self-determination. Viewed in this way, nationalism is the essential equivalent of national territoriality, defined as a political geographic strategy enacted by nationalists seeking to gain control over the fate of the nation by gaining control over or sovereignty in the region identified as their homeland (Kaiser, 1994a).

In multinational, multi-homeland states, nationalism may also be viewed as an interactive process in which the political action programme of the dominant nation triggers rising nationalism among subordinate groups (Hennayake, 1992). This interactive nationalism is clearly apparent in the former USSR, where political, socio-economic and ethnocultural domination by the Russian core heightened separatist tendencies among titular nations in the union republics. The interactive process has continued since independence, as nationalism among the titular nations in the NIS has in turn triggered rising nationalism among subordinate groups. This interactive process has resulted in conflict escalation.

A typology of nationalisms

Nationalism can be broken down into subcategories in order to create a typology of nationalisms. Orridge and Williams (1982, p. 21) in their study of nationalism in Europe made the useful distinction among:

1. "*state nationalism* . . . manifested by groups and regions close to the centre of an old and well-established state";
2. "*unification nationalism* and *irredentism* as attempts to unite culturally similar groups and territories into a larger and more powerful state";
3. "*autonomist nationalism*" as "the claim for some institutional recognition within the larger state"; and
4. "separatist nationalism, the demand for complete independence".

Breuilly (1993) also distinguished between the *unification nationalism* experienced in Germany, Italy and Poland, the *separatist nationalism* that led to the disintegration of the Ottoman and Habsburg empires, and added *anticolonial nationalism* and *subnationalism* to his typology in discussing trends in Asia and Africa. These typologies refer to the different possible power relationships between a nation, its homeland, and the state in which the nation and homeland are located. All of these types of nationalism are present in the former USSR.

Another typology of nationalism could be constructed to refer to specific aspects of national life over which nationalists seek to assert control, which often become activated by specific threats to the nation as they are perceived by its members. *Cultural nationalism* refers to a political action programme which seeks to establish a dominant or hegemonic status for the cultural attributes of the nation (its language, religion, etc.) within the homeland, and frequently arises in reaction to encroachments by the language, religion, etc. of another nation (e.g. russification). *Economic nationalism* refers to a political action programme seeking to ensure that economic activities within the homeland are controlled by the nation, and that ideally members of the nation are the ones to benefit first and foremost from such economic activities. At its extreme, economic nationalism can take the form of protectionism, isolationism or even autarchy, and frequently appears following a period of time in which economic trends and/or decision-making controlled by outsiders have had a negative impact on economic conditions in the homeland. *Demographic nationalism* refers to a political action programme seeking to control the relative demographic weight of national members and non-members in the homeland, usually through policies that discourage in-migration and encourage out-migration of non-members, or that encourage in-migration and higher birth rates among national members. At its extreme, the nationalistic desire for national homogeneity in the homeland can result in "ethnic-cleansing" type programmes including the forcible expulsion of non-members and even their physical extermination (i.e. genocide). This aspect of nationalism also often arises in response to large-scale in-migration of ethnic others, particularly when this migratory trend coincides with a very slowly growing or declining indigenous population (e.g. Latvia and Estonia; Germany and France). Like the first typology of nationalisms, all of these nationalisms are also apparent in the former USSR. We return to these typologies below.

Nations and nationalism in the USSR: alternative interpretations

Two major interpretations of national identity and nationalism are apparent in the literature related to the so-called national question in the USSR, and each interpretation is linked to a specific community of scholars approaching the topic from a particular ideological point of view. *Sovietisation* was the dominant interpretation of trends in national identity in the USSR offered by Soviet politicians and ethnographers, particularly during the 1960s and 1970s. Alternatively, *russification* was the principal interpretation of international trends in the USSR offered by Western analysts (and particularly non-Russian *émigrés*) during the cold war. These clearly identifiable strands began to unravel during the 1980s. A third *nationalisation* interpretation of trends in national identity and nationalism in the USSR is emerging in light of current events and new information about the history of national identity and international relations.

Sovietisation

Sovietisation conformed to the ideological belief inherent within Marxism–Leninism that the nations and nationalities of the state were drawing together into one Soviet people as the USSR moved towards communism. This was based on the belief that nations were a product of capitalism, and that under conditions of developed socialism – in which both class antagonisms and international inequality had been eliminated, and through which individuals were being politically socialised towards an international communist community of interest and belonging – the primacy of national identity was giving way to an international or anational Soviet identity. This is not to say that these politicians and ethnographers saw the demise of nations and their replacement by an international Soviet people as an immediate probability, but rather that the trend in national processes was in the direction of sovietisation. Individuals would likely retain their sense of belonging to distinct ethnocultural communities, but these communities would not serve as politicised interest groups that could compete with the Soviet people for the loyalty of members. Viewed in this way, sovietisation would not deprive the population of its ethnocultural roots, but would reorient them towards an internationalist, as opposed to a nationalist, future.

Data used to support the conclusion that sovietisation was occurring included the increasing number of non-Russians who were becoming fluent in the Russian language, which was treated as the language of international communication; the rising rate of international marriages, which was resulting in an internationalisation of the family and particularly of the children from such international families; migration trends which showed that members of indigenous nations were moving to a more international urban environment; and international attitudinal surveys designed to show that attitudes towards an international work and family environment were

becoming more positive over time. In addition, data indicating that national rates of urbanisation, education and occupational mobility were converging (i.e. that international equalisation across these measures of social mobilisation was occurring) were also used as evidence that the national problem was being solved, since that problem was directly related by Lenin to a history of internal colonialism or international inequality in Russia (e.g. Bromley, 1977, 1984; Grigulevich and Kozlov, 1981; Salikov *et al.*, 1987).

During the 1960s and 1970s a group of Soviet scholars even argued that the federal structure of the state – which had been constructed around ethnonational homelands in order to provide for the political–juridical equality of the nations in the USSR – should be eliminated from the new constitution, because it had fulfilled its mission (i.e. the nations were equals, the national problem had been overcome, and there was no need to retain the federal structure of the state any longer) (Hodnett, 1967; Gleason, 1990a; Kaiser, 1991). Although this viewpoint did not win out, even those who argued in favour of the retention of the federal structure agreed that the nations were fast becoming equalised, and that the federal system was fast outliving its utility in a more equalised and sovietised state where people increasingly viewed the entire USSR as their homeland.

A variation on this theme emerged during the 1970s and developed more fully in the 1980s. While the same interpretation was retained (i.e. that international equalisation had solved the core national problem of inequality that the USSR had inherited), an argument emerged that smaller, more localised national problems would continue to develop along with the development of society and the tendency towards greater international interaction (e.g. Arutyunyan and Bromley, 1986). However, the emergence of these small national problems was generally seen as a result of insensitivity on the part of a few deviant individuals, and did not alter the general conclusion that the national problem had been solved, and that the nations of the USSR were drawing together into one Soviet people.

Gorbachev was a firm believer in the sovietisation of the population, and the reforms initiated during his tenure in office assumed the existence of a Soviet people which would use its new-found freedoms of expression through *glasnost* and democratisation to revitalise the socialist economy and move the USSR forward on the path to communism (Kaiser, 1991). Of course, this assumption of successful sovietisation was one of the major errors made by Gorbachev. The intelligentsia in several of the union republics created organisations nominally in support of *perestroika* (e.g. Sajudis, RUKH), but quickly re-oriented these organisations around the goals of reawakening their nations and mobilising the national members – first in support of an auto-nomist nationalism within a more confederal union, and later in support of separatist nationalism.

The failure of sovietisation to overcome nationalism in the USSR does not mean that there is nothing to the Soviet socialist idea. For example, it has become evident that the majority of the population of the former USSR favours democratic socialism or socialist democracy, and that Sweden rather than the United States is seen as a more ideal state system (Gudkov, 1990). In addition, surveys seeking to explore Russian national self-consciousness have indicated that many Russians (in one survey as many as 42% of respondents) consider themselves more Soviet than Russian (Solchanyk, 1992, p. 34), and a ma-jority of Russians in surveys conducted during the 1980s identified their homeland as the entire USSR, not as Russia (Drobizheva, 1991, p. 5). Even after the disintegration of the USSR, sur-veys indicate that 50% of Russians in Tashkent and 52% of Russians in the cities of Kyrgyzstan still consider the former Soviet Union to be their homeland, compared to only 15% in Tashkent and 12% in Kyrgyzstan who say that Russia is their *rodina* (Ginzburg *et al.*, 1993, p. 89). This is an indicator that sovietisation, with its emphasis on Russians as the most loyal core group in the state, as well as its reliance on the Russian lan-guage as a vehicle for sovietisation, was most successful among Russians. However, the very reliance on the Russians and their language made sovietisation appear not as an international or anational process, but rather as a form of state nationalism, whose end product would look more like a Russian nation-state than an anational Soviet socialist state. Russians obviously perceived this as a much more accept-able future than did non-Russians. Nevertheless, Russians did not act as a Soviet people to prevent the disintegration of the USSR, and in a number of republics Russians clearly supported the independence movements. The degree to which a Soviet identity had attained primacy even among Russians living outside Russia should not be overstated.

Russification

During the cold war, a number of Western ana-lysts (including non-Russian *émigrés* from the USSR) used the same data that Soviet analysts cited as evidence of sovietisation in order to as-sert that the USSR was forcibly assimilating the non-Russians (i.e. engaged in a policy of rus-sification). Linguistic russification, intermarriage with Russians and psychological assimilation to the Russian nation were seen as products of coer-cive nationality policies conducted by a total-itarian centre against its unwilling minority subjects. This russification of the population was frequently depicted in extremely negative terms, as an evil policy conducted by an evil empire against the wishes and interests of the non-Russian population. The USSR was viewed as a multinational empire seeking to become a Rus-sian nation-state at the expense of the non-Russians (e.g. Barghoorn, 1956; Meyer, 1965, pp. 442–7; Rostow, 1967, pp. 211–17; Dzyuba, 1970; Carrere d'Encausse, 1978).

In this interpretation, national identity was often equated with the objective characteristics of the nation (i.e. language, religion, etc.). Linguistic assimilation to the Russian language was fre-quently viewed as the essential equivalent of psychological assimilation. Beyond linguistic russification, Moscow was also charged with the forcible migration of peoples from their home-lands to more russified environments, in which they could be more easily assimilated. The de-portation of nations during the 1940s was viewed not as a unique occurrence, but rather as a more extreme variant of Soviet migration policy. Rus-sian out-migration to the non-Russian home-lands was similarly viewed as the direct result of migration policies designed to dilute the demographic strength of the indigenous nations, to provide the centre with a potential fifth column, and to help spread the russification process more rapidly in the non-Russian areas.

This russification interpretation did differ from the sovietisation viewpoint in its assessment of socio-economic trends. Where Soviet analysts saw equalisation, and interpreted this as an indicator that the national problem had been solved, Western analysts refuted the conclusion that international equalisation was occurring. The USSR was viewed as an empire whose core area exploited the non-Russian periphery for the benefit of the "mother country" (i.e. Russia west of the Urals). Ironically, Lenin's thesis that national problems existed due to internal colonialism also appears to have been shared by Western analysts, who concluded that the national problem existed in the USSR due to the continuation of interregional and international inequality.

The 70-year period of the Soviet Union's existence was viewed as a time of unrelenting attacks against the non-Russian nations by the russifying core. Since the late 1980s, non-Russian nationalists within the former USSR have adopted a similar interpretation of the Soviet period: that their nations were fully formed prior to 1917 (or prior to the time of their incorporation into the Russian Empire), that they experienced a period of dormancy over the past 70 years as they struggled to withstand the denationalisation campaigns emanating from Moscow, and that they have now reawakened to fulfil their destiny. This has been referred to as the "sleeping beauty" analogy (Suny, 1992, p. 24).

As part of the russification interpretation, the recent rise in non-Russian nationalism has been treated as a type of anticolonial nationalism against an imperialistic Russian centre (Carrere d'Encausse, 1978; Conquest, 1986; Motyl, 1992). Non-Russian nationalists from the most developed and least developed republics in the state complained of colonial exploitation by Moscow and argued in favour of "decolonisation". In Central Asia, nationalists charged the centre with economic exploitation, and with converting the region into a cotton colony (Gleason, 1990b). While there is some validity to this charge, it is also true that the central government sent more

investment funds to Central Asia than it extracted throughout most of the post-war period. In the Baltic republics, which had the highest standard of living in the USSR, the argument of economic exploitation centred around the idea that Moscow extracted more than it returned (even though the economic data did not support this conclusion), and that the Baltic states would be much more developed if only they had retained their independence following the Second World War. Finland was pointed to as proof of this conclusion, since Finland had gained its independence from Russia at the same time as the Baltic states but had retained its independence after the Second World War (Taagepera, 1989; Smulders, 1990; Bond and Sagers, 1990; McAuley, 1991).

The russification interpretation of national processes is certainly questionable, as is the view of the USSR as an imperialistic, totalitarian state. First, little linguistic russification occurred during the era of "developed socialism", and the rate of increase was declining over time (Table 2.1). Members of titular nations also intermarried infrequently, and the children of such international families most often adopted the national identity of the indigenous nation, not Russian. Beyond this, the russification that did occur resulted not in the demise of nations, but in rising anti-Moscow and anti-Russian nationalism. In Latvia and Estonia, for example, the in-migration of Russians (i.e. demographic russification) coincided with rising nationalism, not with the russification of the indigenous populations and their loss of national identity. Similarly, linguistic russification in several of the union republics proved counter-productive to the goal of assimilation, since it served to raise national self-consciousness among non-Russians that was both anti-Moscow and anti-Russian. One of the earliest phases of rising nationalism during the late 1980s was cultural nationalism, as titular nationalists passed language laws declaring the language of the titular nation to be the official language of the republic. In places where Russian in-migration had been relatively high, the timing of the rise in cultural nationalism coincided with demographic nationalism (e.g. Estonia and Latvia). In addition, feelings of internal colonialism encouraged a rise in economic nationalism, which paradoxically was most pronounced in the most, not the least, developed republics.

Table 2.1 *Linguistic russification in the USSR, 1959–89 (% claiming Russian as the first language, and percentage point change)*

Nation	Home-land 1959	Urban 1989	% change	Home-land 1959	Rural 1989	% change	Out-side 1959	Urban 1989	% change	Out-side 1959	Rural 1989	% change
Ukrainian	15	19	4	1	2	1	54	57	3	42	50	8
Belarusian	22	30	8	1	3	2	61	64	−3	46	52	6
Moldovan	9	11	2	1	1	0	36	36	0	8	11	3
Uzbek	1	1	0	0	0	0	3	5	2	0	1	1
Kazakh	2	3	1	0	1	1	7	9	2	2	4	2
Kyrgyz	1	1	0	0	0	0	7	7	0	0	1	1
Tajik	2	2	0	0	0	0	3	4	1	0	1	1
Turkmen	2	2	0	0	0	0	12	13	1	1	1	0
Georgian	1	0	−1	0	0	0	31	33	2	12	17	5
Azeri	2	1	−1	0	0	0	11	16	5	1	3	2
Armenian[a]	1	0	−1	0	0	0	27	30	3	5	9	4
Lithuanian	0	0	0	0	0	0	21	38	17	9	23	14
Latvian	2	3	1	1	1	0	54	56	2	31	43	12
Estonian	1	2	1	0	0	0	55	61	6	29	42	13
Groups whose homelands are:[b]												
ASSRs	9	13	4	0	1	1	23	30	7	8	15	7
Autonomous oblasts	14	9	−5	3	3	0	34	33	−1	14	21	7
Autonomous okrugs	29	49	20	12	26	14	40	53	13	18	32	14

[a] Armenians outside the homeland exclude those in Nagorno-Karabakh autonomous *oblast*. The figures for Armenians in Nagorno-Karabakh are: urban: 1959 = 7%, 1989 = 3%; rural: 1959 = 1%, 1989 = 0%.
[b] Averages are unweighted, with each national community treated as one observation.
Source: Kaiser (1994a, pp. 276–81).

The trends during the 1980s regarding the "national question" indicate that something other than (or at least in addition to) sovietisation and russification was occurring in the USSR. An alternative interpretation of trends in national identity and nationalism is that non-Russians with homelands in the USSR were experiencing a growing sense of national self-consciousness during the Soviet period, and that nationalisation rather than russification or sovietisation had become the dominant trend by the end of the Stalin era. The main elements of this nationalisation process are briefly discussed below.

Nationalisation

As an alternative to the sovietisation and russification interpretations, nationalisation refers to a process through which national identity

became mass-based in the USSR, and nationalism became established as the dominant ideology within the state. Unlike the previous two interpretations, the nationalisation interpretation views the Soviet period as primarily one of nation-making, not nation-destroying. The pre-Soviet historical record indicates that for the most part nations as mass-based communities of interest and belonging did not exist prior to 1917. Most of the rural, illiterate and relatively immobile peasantry lived highly localised lives, and a localist mentality continued to guide the way people viewed themselves, outsiders, and the world in which they lived. A romantic nationalism did rise among indigenous élites during the nineteenth century, but these élites and their nationalistic message were not welcomed with open arms in the rural countryside, where urban intellectuals from outside the local village were treated with a good deal of suspicion and scepticism.

As the rural peasantry became more geographically and socially mobilised towards the end of the nineteenth century, localism began to give way to nationalism, and home began to be redefined as more expansive national homelands, rather than – or in addition to – the local village. The vernacularisation and standardisation of written languages, along with increasing literacy and education, also contributed to this nationalisation process, but this ethnolinguistic consolidation was also only beginning to develop by the end of the nineteenth century.

The process of nationalisation had proceeded furthest and begun to assume a mass basis in the Baltic *guberniyas*, but had not become much more than an ideology of the intelligentsia outside the north-western regions of the Russian Empire. During the late nineteenth century, the more nationalised regions were targeted for russification by the Tsarist government. This policy proved counter-productive, and autonomist and separatist nationalism rose in response to the increasingly Russocentric core.

After 1917, mass-based nationalisation among non-Russians in the USSR continued to be the major trend. The nationalisation of the masses was enhanced not only by the rapid socio-economic development in the state, but also by the federalisation of the state along national territorial lines and indigenisation (*korenizatsiya*) policies designed to accelerate the process of international equalisation by targeting members of indigenous groups for preferential treatment. First, rapid industrialisation and the collectivisation of agriculture brought revolutionary changes which tended to accelerate the breakdown of localism. At the same time, the massive disruption in people's lives caused by these socio-economic upheavals tended to orient them towards a national, rather than an international or anational, Soviet identity.

Second, the federalisation of the state structure along national–territorial lines, created to provide for the political–juridical equality of non-Russian nations and thus indirectly to promote sovietisation (and also to help put an end to the civil war and assist in the reincorporation of territories lost between 1917 and 1921), helped give form to national homelands that for the rural masses of many groups in the state were only vaguely defined. Home was still defined in highly localised terms, or if this localised sense of homeland had been eroded, a new national homeland had not yet become clearly delimited in the minds of the national membership. From the 1920s forward, although nationalists frequently continued to claim more than their home republics as their rightful homelands, the officially delimited territories came to be seen as the geographic minima of national homelands. The federalisation of the state along national territorial lines was not a sham in this regard, because it promoted the territorialisation and politicisation of indigenous groups, and was thus a crucial element in the nationalisation process. The lack of political power in the republics, and particularly the contrast between paper rights and actual power relations between Moscow and the republics, also tended to heighten sentiments in favour of autonomist and separatist nationalism.

Third, the state encouraged nationalisation through indigenisation (*korenizatsiya*) policies during the 1920s and early 1930s. Members of the indigenous groups were given preferential treatment in access to higher education, higher status employment and élite political positions in their homelands. Literary languages for titular groups were standardised or created where none existed, and education – including adult literacy programmes – was conducted in these languages. *Korenizatsiya* accelerated the nationalisation of the indigenous masses, and because this policy was geographically restricted to the home republics, it also encouraged the development of a nationalistic sense of exclusiveness among members of the titular groups towards their homelands.

These socio-economic trends and national policies were not resulting in the sovietisation of the population (as they were theoretically designed to do), and during the late 1920s and early 1930s it became increasingly apparent that nationalism rather than internationalism was developing among the indigenes in each home republic within the state. At this time, a reorientation of national polices in favour of russification was enacted, and indigenous political and economic élites were purged. However, this radical policy shift – coming as it did on the heels of policies promoting mass-based nationalisation – proved counter-productive. The shift in favour of russification and against what was termed "local nationalism" during the late Stalin era did drive nationalism as an overt political action programme underground, but this policy reorientation did not enhance the prospects for either a sovietisation or a russification of the population.

During the post-Stalin period, the Russian nation and language retained their dominant status in Soviet society. Every political leader from Khrushchev to Gorbachev also asserted that the national problem had been solved through the equalisation of nations in the USSR. Soviet ethnographers such as Yulian Bromley asserted that the process of national consolidation had been completed, and that the drawing together of nations into one Soviet people was becoming the dominant trend in the USSR (Bromley, 1983). Other aspects of this sovietisation thesis have been noted above.

Given the fact that the Russian language had been firmly established as the USSR's lingua franca by the 1950s, it is surprising how little linguistic russification of the population actually took place between 1959 and 1989 (Table 2.1). Most of this linguistic russification took place among individuals living outside their home republics. In the homeland, except for groups with homelands at the autonomous-*okrug* level in the federal hierarchy, almost no linguistic russification occurred between 1959 and 1989 according to census results.

Non-Russians were becoming more fluent in the Russian language, but typically allocated it to the status of second language, and reserved for their national languages the status of first language. This was not necessarily a ranking that accorded with fluency levels; studies of the gap between fluency and the ranking accorded the indigenous and Russian languages indicated that particularly in urban areas, members of the titular nations who were more fluent in Russian none the less tended to identify the indigenous language as their first language (Snezhkova, 1982, p. 86; Guboglo, 1984, p. 125; Silver, 1986; Kaiser, 1994a, p. 283).

Growing bilingualism and fluency in Russian were cited as indicators of the drawing together of nations (Guboglo, 1969, p. 17). However, another interpretation of these data is that members of the titular nations in their homelands resented the need to learn Russian, which had become the language of upward mobility. These indigenes used the censuses to vote for their indigenous languages by naming them first, and allocating Russian to secondary status. Far from reflecting a drawing together of nations, these data appear to reflect a growing cultural nationalism among the titular nations. This indicates that even the limited russification that did occur tended to heighten national consciousness and was thus counter-productive to the assimilative goals of the state.

Beyond the language data, intermarriage trends also indicated that something other than sovietisation or russification was occurring. First, as in the case of the language data, members of the titular nations tended to intermarry at very low rates in both urban and rural areas of their homelands. More importantly, in a majority of those cases where members of the indigenous nations did intermarry, the children of these international couples most often chose the national identity of the indigenous group. This held true even when the other parent was a Russian. It was only among non-indigenes that intermarriage was high, and the children of these international families did adopt a Russian national identity more frequently when one parent was a Russian, or an indigenous national identity when one parent was a member of the titular nation. Overall, nationalisation was the dominant trend, exceeding both internationalisation (i.e. sovietisation) and russification (Kaiser, 1994a, pp. 295–321).

In the socio-economic sphere, indigenisation began again in the post-Stalin era. However, unlike the *korenizatsiya* programmes of the 1920s and early 1930s, the indigenisation that occurred between 1959 and 1989 was increasingly part of a nationalisation programme initiated not at the centre from above, but within the republics themselves, from below. Indigenous élites were increasingly using their influence to promote members of their own nations, and this indigenisation from below supplemented indigenisation policies from above. As a result of these indigenisation processes, by the 1980s members of the indigenous nations had become relatively over-represented in higher education, high status

Table 2.2 *Indigenous proportions of total population, students entering higher education, directors of economic enterprises and political representatives, 1989 (%)*

Nation	% of total population	% of students	% of directors	% of political representatives in:		
				CPD[a]	RSS[b]	LS[c]
Russian	82	80	77	71	78	—
Ukrainian	73	67	79	69	75	86
Belarusian	78	71	78	70	74	86
Moldovan	64	71	50	72	69	77
Uzbek	71	71	68	70	78	79
Kazakh	40	54	40	40	54	54
Kyrgyz	52	65	55	59	64	69
Tajik	62	63	66	67	75	69
Turkmen	72	78	72	76	74	81
Georgian	70	89	89	73	—	—
Azeri	83	91	94	76	—	—
Armenian	93	99	99	88	—	—
Lithuanian	80	88	92	85	87	—
Latvian	52	54	63	78	70	83
Estonian	62	78	82	81	77	—

[a] Congress of People's Deputies.
[b] Republican supreme soviets.
[c] Local soviets.
Source: Kaiser (1994a, pp. 233, 241, 349).

employment and élite political positions in their homelands (Table 2.2).

The proportion of indigenous students entering higher education during the 1980s equalled or exceeded the indigenous proportion of the total population in all but the three Slavic republics. The proportion of indigenous directors of economic enterprises also equalled or exceeded the indigenous proportion of the total population in all union republics except Russia, Moldova and Uzbekistan. The proportion of indigenes elected to the first Congress of People's Deputies also equalled or exceeded the indigenous proportion of the total population in every republic except the Slavic republics, Uzbekistan, Azerbaijan and Armenia. Within the republics, the indigenous proportion of seats in the republican supreme soviets and local soviets exceeded the indigenous proportion of the total population in all republics for which data are available except Russia and Belarus. It is noteworthy that the one nation that shows up in every category as relatively under-represented in its own homeland is the Russian nation, and this is certainly counter-intuitive given the status of the

Russians as the dominant nation in the state. The USSR hardly appeared to be a Russocentric empire, at least by the 1980s. Rather, it had become a multinational state in which a two-tiered system of ethnic stratification existed, with Russians in a dominant position in the USSR as a whole, but also with each indigenous nation occupying a privileged standing within its own home republic.

On the basis of these and other data, it is more reasonable to conclude that members of the indigenous nations were not only becoming more nationally self-conscious, but were also becoming more nationalistic regarding their homelands. International equalisation, to the extent that it was occurring in the USSR, was not resulting in the sovietisation or russification of the population, but rather in rising indigenous nationalism as socially mobilised members of the titular nations competed for dominance in their homelands. International equalisation in the country as a whole was occurring as a result of rising indigenous dominance in each union republic, and was thus unlikely to promote an international coming together of the population.

Soviet ethnographers conducting sociological surveys during the mid-1980s were coming to the same conclusions (e.g. Arutyunyan and Bromley, 1986). The national problem confronting the USSR became more intractable as nationalisation proceeded, so that at the time of the reforms initiated under Gorbachev, separatist nationalism rather than an internationalist revitalisation of Soviet society became the dominant ideology and political action programme in society.

A typology of nationalisms in the newly independent states

Viewing the history of national identity and nationalism in the USSR in this way, it becomes apparent that *autonomist* and *separatist nationalism* were ongoing political action programmes in the union republics throughout much of the post-war period, rather than new political geographic projects that began with the reforms initiated by Gorbachev. For non-Russians in the state, what began as a push by nationalists for greater territorial autonomy in their homelands in the cultural and socio-demographic sectors (i.e. a cultural and demographic nationalism), expanded to include demands for greater decentralisation of economic and political decision-making for nations in their homelands during the 1980s. Demands for greater autonomy in a more confederal union were not satisfied by Moscow, though they were partially met. This led to an escalation of demands in favour of separation, particularly in the Baltic republics, in Moldova and in Georgia. Throughout this period, Gorbachev refused to believe that the indigenous masses favoured separation, and preferred to see the rising demands for independence as the work of a few nationalist deviants in a sea of Soviet people (e.g. Gorbachev, 1989, p. 13). For the political centre steeped in the myths of successful sovietisation, and for a leader who was successfully sovietised, it was impossible to accept the reality of mass-based nationalism in the republics.

The separatist nationalism in the Baltics, Moldova and Georgia diffused to the indigenous political élites in other union republics in the state, the most important of which was Ukraine.

The Armenian leadership became more independence-oriented when Moscow refused to side with it in the Nagorno-Karabakh dispute; Azerbaijani political élites also made the transition from autonomist to separatist nationalism after the Soviet army was called into Baku to quell disturbances related to Nagorno-Karabakh. Indigenous political élites in the other republics (Belarus, Central Asia and Kazakhstan) remained more autonomist in their orientation, preferring a renewal of the union on a more confederal basis over outright independence. The political élites of those groups with autonomous republics also began to demand greater autonomy in a more decentralised Russia, or outright independence from the union republic in which their autonomous territories were located (e.g. Abkhazia and South Ossetia in Georgia).

By 1990, Gorbachev had accepted the reality of separatist nationalism, and attempted to regain control over the processes of disintegration through a law on secession and by redrafting the 1992 Union Treaty (*Pravda*, 7.4.90, p. 2; Goble, 1990; Kaiser, 1994a, pp. 345–6). The law on secession made it virtually impossible for a republic to secede from the union, and was ignored by nationalistic indigenous élites in the union republics. The early drafts of the new union treaty were also rejected, both because they were not confederal enough (i.e. they reserved too much political power for the centre), and because they were written by political élites in Moscow. If the new union was to be a confederation, so the political élites in the union republics argued, then they – not the centre – should be the ones to create the new union treaty. Finally, in April 1991 at a meeting between the political leaders of nine union republics (the Slavic republics, Central Asia and Kazakhstan, and Azerbaijan) and Gorbachev (the so-called nine plus one agreements), a confederal union treaty was drafted to create a "Union of Sovereign States" which met the demands of the republic political élites who participated. The new union treaty was to be signed in August of that year, and hardliners launched the August 1991 coup attempt in an effort to prevent this from happening. The failed coup changed the political climate in the union republics in favour of outright independence, and was the final event leading to the collapse of the state and the Communist Party.

State nationalism in the newly independent states (NIS)

Since the end of 1991, the dominant political action programme in the NIS has been a form of *state nationalism*, through which indigenous nationalist élites are attempting to restructure society with the goal of establishing the cultural, economic and political hegemony of the titular nation, and to exclude or subordinate members of non-titular nations living there. This nation-state building project first became apparent in the declarations of sovereignty and independence, which were for the most part made in the name of the titular nations, and not in the name of the entire population of the republic. Nearly all of the new constitutions also begin their preambles with a statement to the effect that independence has been declared in order to secure the fate of the titular nation, which has reawakened to fulfil its destiny. This constitutional nationalism, even in constitutions which otherwise are not centred around the goal of titular nation hegemony, sets a tone of international differentiation and titular favouritism that have added to feelings of alienation among members of non-titular groups.

Constitutional nationalism has been supplemented by language laws in every state, and by citizenship, immigration and property laws in selective states in a more or less concerted effort to deconstruct the old two-tiered ethnic stratification system which placed Russians in a dominant position throughout the USSR and also gave preferential treatment to members of the titular nations living in their homelands, and to reconstruct a new ethnic stratification system in which members of the titular nations alone hold a dominant or hegemonic position in their homelands (Chinn and Kaiser, 1996).

Regional variations do exist in the degree to which this nation-state building project dominates political decision-making. It is clearly more prominent in Latvia, Estonia and Georgia, and least apparent in Belarus, the Russian Federation and Central Asia. Nevertheless, it does appear that even in those republics where members of the titular nations were not strongly in favour of independence and which did not begin their independent lives preoccupied with this political action programme, a reorientation in favour of

state nationalism is occurring. For example, political élites in Kazakhstan have clearly shifted from a more ethnically neutral state-building programme at the time of independence towards a nation-state building project since 1993. This shift in favour of state nationalism and the politics of ethnic exclusion are evident in the new constitution, in the citizenship law which extends citizenship to Kazakhs living outside the state and provides funds to help pay for their return, but does not allow Russians to hold dual citizenship in Kazakhstan and Russia. It is also evident in the creation of new Kazakh schools that not only teach in the Kazakh language, but are oriented around a Kazakh cultural emersion programme, and also in the growing favouritism shown Kazakhs in access to higher education. Kazakh favouritism is also apparent in the hiring of economic and political élites, and is resulting in a kazakhisation of the upper strata of society not only in southern Kazakhstan, where Kazakhs are in the majority, but also in northern Kazakhstan, where Russians comprise the largest segment of the population. Finally, the recent elections of March 1994 were manipulated to ensure that the new parliament would be overwhelmingly Kazakh in ethnic composition, and that it would be compliant with Nazarbayev and with this new nation-state building project (Kaiser and Chinn, 1995).

In Russia, it is also apparent that the central government which began with a more populist than nationalist political action programme, has shifted towards state nationalism (though not yet a nation-state building project). Although the new Russian constitution is the only one that opens its preamble with the phrase "We, the multinational people of . . .", it none the less represents a significant setback for non-Russians in the state, and particularly for those with homelands there. Earlier drafts of the Russian Constitution and the Federation Treaty treated the autonomous republics as sovereign entities within the Russian Federation, and the federation itself was an ethnically based federation. However, in the constitution adopted in December 1993, the federation includes not only the non-Russian republics in the state, but treats the Russian *oblasts* as the essential equivalents of the non-Russian republics, greatly diminishing the bargaining power of the republics at the centre. In addition, the word "sovereignty" has been removed from the description of the republics' status in the state, and has been replaced by "local self-government", which Russian *oblasts* also exercise. Finally, the right to secession and the text of the Federation Treaty, included in previous drafts of the Russian Constitution, were removed from the final version adopted (Administratsii Prezidenta Rossiyskoy Federatsii, 1993).

The only states where political élites do not appear to be engaged in some degree of nation-state building are Tajikistan and Belarus. In Tajikistan, state nationalism was the dominant political action programme in 1992 when nationalists gained control in Dushanbe, but this ended with their ouster by the end of that year. Since 1992, divisive *subnationalism* has been more apparent, as ethnoregional divisions within the Tajik nation have warred on one another, indicating that a Tajik national identity is relatively weak. Belarusian national self-consciousness is also relatively weak, and this is reflected in the limited degree to which state nationalism has emerged as a dominant political action programme. Indeed, reunification with the Russian Federation in a new confederal relationship appears to be favoured over nation-state building by a majority of Belarusians.

In response to the state nationalism of the titular nationalist élites, members of non-titular groups have also become more nationalistic, and depending on their political power status, have begun to organise politically around one of the types of nationalism cited above. As an alternative to political organisation *in situ*, members of non-titular groups have also been emigrating to their respective homelands in increasing numbers. These trends are resulting both in the increasing segregation of national communities in regions identified by members as ancestral homelands, and also in rising tensions and conflicts between communities whose territorial claims overlap. The nationalisation of space in the former USSR, and the divisive nature of rising nationalism, have reduced the prospects for reintegration and conflict management, although there are selective cases of success in both these regards.

Unification nationalism/irredentism

In the Baltic states, there is the potential for rising irredentism among Russians, particularly those living in north-eastern Estonia. The territorial autonomy referendums held in Narva and Sillamae could be a harbinger of things to come if Estonia continues to define itself in exclusionary nationalist terms. However, these referendums appear to have been more an effort by Russians who had been excluded from citizenship and political representation in Estonia to get Tallinn's attention, rather than a serious effort to secure territorial autonomy in north-eastern Estonia. Other irredentist claims in the region – by Estonia and Latvia for border regions lost to the Russian republic following their incorporation into the USSR, by Lithuanian nationalists for Kaliningrad *oblast*, and by Belarusian nationalists for border regions in Lithuania (Kolossov, 1992, pp. 42–3) – do not appear to be claims that are capable of mobilising mass support.

In Ukraine, several potentially serious irredentist claims exist, and have strained interstate relations between Ukraine and Russia, and between Ukraine and Romania. The most serious case of irredentism is in Crimea, which was presented as a gift to Ukraine by Russia in 1954 to mark the 300th anniversary of Russian–Ukrainian ties. None the less, Crimea remained a Russian *oblast* in national orientation as well as demographically. When Ukraine declared its sovereignty, Crimea passed a resolution to declare its own sovereignty as an autonomous republic, in order to be in a position to secede from Ukraine and join Russia if Ukraine became independent. The recent election of a Russian nationalist as president of Crimea, the decision to hold a referendum on the fate of Crimea on 27 March (which was declared illegal in Kiev), and the declaration by the Sevastopol city government that Sevastopol places itself under Russia's jurisdiction indicate that irredentism is a strong sentiment among the Russians of Crimea (Chinn and Kaiser, 1996). Kiev has offered limited territorial autonomy for Crimea; it remains to be seen whether or not this will be sufficient to satisfy Russian nationalism in the region.

Russians are also concentrated in eastern Ukraine, and the potential for irredentism exists among this population as well. At present, they appear to be pushing for a federal Ukraine in which they will have greater autonomy (i.e. an autonomist nationalism agenda) (Volkov, 1993), but if this is unsuccessful, they may push more strongly for outright independence from Ukraine and admission into Russia. At present, Russia makes no official claim to these regions, but does see itself as the defender of Russians in the so-called "near abroad".

Romanian nationalists in Romania do claim parts of the western *oblasts* of Ukraine that historically belonged to Romania. However, as there are few Romanians currently living in Odessa *oblast* and the former northern Bukovina, mass-based sentiments in favour of irredentism here do not appear likely. Finally, in Zakarpatskaya *oblast*, which was part of Czechoslovakia prior to the Second World War, a rising Ruthenian nationalism is evident. However, the dominant political action programme called for by Ruthenian nationalists here is not reunification with Slovakia, but federalisation of Ukraine and territorial autonomy for "Ruthenia" (i.e. autonomist nationalism).

In Moldova, the early nationalist political agenda in the republic viewed reunification with Romania as the ultimate objective, and Romania supports this goal. However, in recent elections the nationalist party running on the reunification platform lost a great deal of support, and a Moldovan nation-state building project is more popular at present. This raises serious questions regarding assumptions made that a Moldovan national identity does not exist and that Moldovans are members of the Romanian nation. The behaviour of Moldovans since independence appears to provide substantial evidence that a separate mass-based national self-consciousness has emerged among the Moldovans during the past half-century (King, 1994).

The early support for reunification with Romania among Moldovan political élites gave rise to irredentism among Russians and Ukrainians in Transdniestria, where a declaration of sovereignty was made in 1991. It is unclear what impact the new state-nationalist orientation among Moldovan political élites will have on Russian and Ukrainian irredentism in Transdniestria, although the recent agreement by

President Snegur to allow for territorial autonomy in this region of the state may satisfy Russian and Ukrainian nationalists (Chinn and Kaiser, 1996).

In Russia, there are nationalist forces seeking to gather in the Russian regions of the border states. From Solzhenitsyn to Zhirinovsky, Russian nationalists have tended to define Russia in geographic terms broader than the Russian republic. For Solzhenitsyn (1990, p. 2), Russia must at least include Belarus, Ukraine and northern Kazakhstan. For Zhirinovsky, Russia is defined as the Russian Empire at its greatest geographic expanse, including not only all of the territory of the former USSR, but also Congress Poland and Finland (Kipp, 1994, pp. 76–8). While the latter's extremism may not need to be taken seriously, it is important to recall that Russians have tended to define home as all of the USSR, and not just as the Russian Federation. A strong sentiment exists that the former union republics are regions for which Russia has a special interest and responsibility, which is implied in the term "near abroad" and in arguments made by Yeltsin, Kozyrev and Grachev about the need to defend Russian interests not within the borders of Russia but on the external borders of the former USSR (Foye, 1993, 1994; Crow, 1993; FBIS, 1993). If more nationalistic forces come to power in Russia, or if the Yeltsin government continues moving towards a more nationalistic orientation, and if the status of Russians outside Russia continues to deteriorate, Russia is certain to become more oriented towards unification nationalism. This represents the greatest potential danger in the former Soviet Union, which is of particular concern in Ukraine and Kazakhstan.

Other forms of unification nationalism within Russia itself are apparent. First, and as mentioned above, the shift in favour of state nationalism in Russia has diminished the territorial autonomy of the non-Russian republics. The tendency of Russian élites to see the ethnically based federalisation of the state as a problem whose solution is the dismantling of the non-Russian republics, clearly corresponds to the political objectives of unification nationalism (i.e. of restructuring Russia as a unitary nation-state). If Moscow attempts to fulfil this goal, it will almost certainly bring about the result that Russian nationalists fear most – the disintegration of "Mother Russia".

Several non-Russian nations with homelands in Russia have also been engaged in unification nationalism/irredentism. In North Caucasia, several such movements exist. Ossetian nationalists in South Ossetia, reacting to the state nationalism of Georgians under the late Gamsakhurdia, declared their intent to secede from the state and to unite with North Ossetia in the Russian Federation. Georgian nationalist political élites responded to this irredentism by abolishing the South Ossetian autonomous *oblast*, and the conflict between Georgians and Ossetians escalated to civil war in 1992. An unstable truce exists at present, but the basic ethnoterritorial dispute remains unresolved.

Irredentist claims have also been made by the Ingush to a region of North Ossetia that belonged to the Ingush republic prior to its abolition and the deportation of the Ingush in 1944. Attempts by the Ingush to move into North Ossetia have sparked conflict there as well. In addition to these two cases, an overarching unification nationalism binding the indigenes of the region into a Confederation of Caucasus Peoples has also been a political action programme in the region, though the degree to which such a confederation has mass-based support among the indigenous nations in the region is an open question.

Outside of North Caucasia, Kazan Tatar nationalists have also called for the unification of Tatarstan and Bashkortostan, based on the idea that Bashkirs are ethnically related to Tatars, and also that more Tatars than Bashkirs live in Bashkortostan (Kolossov, 1992, pp. 42–7). Nor surprisingly, the Bashkirs do not agree that their nation and homeland should merge with Tatarstan. Finally, the Buryats, who have three official autonomous units around Lake Baykal (Republic of Buryatia, Agin Buryat autonomous *okrug*, and Ust'-Ordin Buryat autonomous *okrug*), have called for the reunification of these units, which together comprised the Buryat–Mongol ASSR from 1923 to 1937, but which were separated into three autonomous units after 1937 (Kolossov, 1992, p. 19).

In Transcaucasia, in addition to the case of South Ossetia mentioned above, Georgia also faces irredentism among the Armenians and Azeris living in the southern regions of the state. Nagorno-Karabakh, the bloodiest and longest-lasting conflict in the region and in the former USSR as a whole, is also a case of irredentist nationalism. In 1993, Armenia won a number of military victories in the region between Armenia and Nagorno-Karabakh, and has ethnically cleansed these regions of Azeris, thus paving the way for the unification of Nagorno-Karabakh and the adjacent borderlands with Armenia, which Armenians claim as part of their extended homeland (Yemel'yanenko, 1994). Azeri nationalists continue to claim this region as an integral part of their national homeland, and will continue to seek to retain control over this region. However, the loss of an Azeri demographic presence in the region makes Azerbaijan's hold on it more tenuous.

Lezgins live on both sides of the Azerbaijan–Dagestan border, and the Lezgin nationalist movement Sadval has pressed for the unification of these two parts of the Lezgin nation and homeland (Toshchenko, 1994). Finally, a potential unification nationalism comparable in some ways to Moldova exists between Azerbaijan and the Azeri population living in northern Iran. The reunification of Germany served as a catalyst for reunification calls in Azerbaijan, but this has not since become a dominant political action programme in the state.

In Kazakhstan, two cases of unification/irredentist nationalism exist, although both have been muted to date. First, Kazakhs live in western China along the border with Kazakhstan, and Kazakh nationalists consider this region to be part of their ancestral homeland. However, a unification nationalist project is at present endorsed only by extremists, and does not have the support either of Kazakh political élites or of the Kazakh population generally. Second, Russians who presently comprise 36% of the total population of Kazakhstan, and who represent a majority or plurality of the population in several northern *oblasts* adjacent to Russia have become more irredentist, particularly as the Kazakh political élites have become more oriented towards a nation-state building project in Kazakhstan. At present, Russian separatist movements do exist in northern Kazakhstan, but they have not yet been successful in attaining popular support. However, the recent decision by Nazarbayev to disallow dual citizenship in Kazakhstan will force Russians to choose either a Russian or Kazakhstan citizenship in the near future. Given the Kazakh nationalist reorientation of the state, it would not be at all surprising if a majority of Russians in the north opted for Russian citizenship, and this would result in a *de facto* secession from Kazakhstan (Kaiser and Chinn, 1995).

Finally, in Central Asia, irredentism is for the most part concentrated in the Fergana Valley, where Tajik, Kyrgyz and Uzbek nationalists living in areas outside but adjacent to their titular states have called for the redrawing of borders to incorporate these homeland regions within the national state. Thus far, although ethnoterritorial conflicts have occurred in this region (the most notable of which was in Osh *oblast* between Kyrgyz and Uzbeks), the predominant sentiment appears to be in favour of territorial autonomy rather than outright separation. Out-migration is also occurring (e.g. Gosudarstvennoye Statisticheskoye Agentstvo Kyrgyzskoy Respubliki, 1993, pp. 145–62).

Uzbeks are represented in the greatest numbers outside their home state, and their calls for Uzbek territorial autonomy outside Uzbekistan or for outright separation and accession to Uzbekistan have raised concerns throughout Central Asia that Uzbeks and Uzbekistan are seeking to become the dominant nation and state in the region. Similarly, calls for the re-creation of Turkestan are interpreted by non-Uzbeks as an attempt by Uzbeks to reconstruct Central Asia as Greater Uzbekistan (Kaiser, 1994a, b).

As a third aspect of unification nationalism and irredentism in Central Asia, Tajiks, Uzbeks and Turkmen are found outside Central Asia in northern Afghanistan and Iran, and represent a potential irredentist challenge to interstate borders in the region. The current conflict in Tajikistan has involved Afghanistan, which is itself engaged in an interethnic civil war, and the unification of Tajiks in Afghanistan and Tajikistan into a Greater Tajikistan has been raised as a possible outcome of the current conflict. However, a divisive subnationalism rather than an overarching unification nationalism is the dominant political project of the moment, indicating that

national identity among the Tajiks even in Tajikistan is not strong. The extension of Tajikistan through unification nationalism to a population that has not been part of Tajikistan or the Tajik nation in the former USSR for generations is thus judged to be a remote possibility.

Autonomist–separatist nationalism

As was noted above, unification nationalism and irredentism for one nation is usually separatism for another. In addition to the cases discussed in the previous section, autonomist nationalism has occurred in virtually all the former union republics, and includes not only those nations with officially delimited homelands in the old Soviet federal structure, but also non-titular groups that never received such official status.

Cases of autonomist nationalism not already mentioned include: Poles in Lithuania, Crimean Tatars, a deported nation seeking to reclaim their Crimean homeland, Gagauz in Moldova, Karakalpaks in Uzbekistan, Pamir Tajiks in Tajikistan, and Uygurs in Kazakhstan (the latter of which is also a potential case of irredentism). Nearly all of those indigenous groups with republics in Russia began to demand greater territorial autonomy in response to the rising territorial nationalism in Russia and the other former union republics. Even within some of the autonomous republics, an autonomist and even separatist nationalism is at work. For example, Dagestan, a republic that is home to over a dozen ethnonational communities, is currently experiencing a process of federalisation, as each indigenous group seeks greater autonomy in the region of the republic claimed as its homeland. The Chechen–Ingush republic has already split into Chechnya and Ingushetia, and similar separatism has risen in Kabardino-Balkaria. Autonomist nationalism has been less apparent among those groups in northern Siberia whose homelands are designated as autonomous *okrugs*, and this is a reflection of the limited degree to which national identity and nationalism have attained primacy among these groups. These titular groups – small in number and inhabiting large land areas – were the only groups that experienced substantial assimilation (mostly russification) even in their home regions. In addition, while a degree of autonomism has been apparent among the Finnic groups with homelands in the Volga–Urals and north-west Russia, this has also been muted. These groups, and particularly the Karelians and Mordvins, have also experienced greater levels of assimilation with the Russians (Kaiser, 1994a, p. 322).

Most of the autonomist nationalism has been limited to demands for some degree of territorial autonomy within a more politically decentralised state, and few of these movements have become openly separatist. In addition to those cases cited under unification nationalism/irredentism, separatist nationalism is concentrated in the North Caucasus region. In addition to Abkhazia, where separation from Georgia has been the goal of Abkhazian nationalists since at least the 1970s, Chechnya has also separated from Ingushetia and declared its independence from the Russian Federation. Chechnya did not sign the Federation Treaty, an did not take part in political elections in Russia prior to 1995. An early decision by Yeltsin to send troops to Chechnya was reversed before conflict began, and for the most part Russia chose to ignore Chechnya's declaration of independence, and to consider it part of the Russian Federation (for example, it is listed in the new constitution). However, since late 1994, Russia has actively supported a political organisation opposed to President Dudayev and his separatist nationalism (Wishnevsky, 1994). In December 1994, Russian troops invaded Chechnya and conquered the capital of Grozny. Although mass-based support for General Dudayev's separatist nationalism programme appears to have been limited among Chechens prior to the invasion, the war itself has tended to increase the popular support for outright independence in Chechnya.

The Ingush and Ingushetia also appear to have become more separatist since 1993, in large part due to the failure of Russia to side with them in their territorial dispute with North Ossetia (see above). The Chechens and Ingush decided to separate from one another in large part because the Chechen political élites were much more separatist than their Ingush counterparts. This no longer appears to be the case. Both the Chechens and Ingush were deported during the 1940s, and

this event clearly did not kill these groups, as the title of a book on the subject by Robert Conquest (1970) suggests, but rather served as a catalyst for their nationalisation.

Also in North Caucasus, the Confederation of Mountain Peoples may serve as a geopolitical foundation for the eventual secession of this region from Russia. However, at this point the confederation is part of Russia and its members have not proclaimed separation as their political objective.

Outside of North Caucasia, the Kazan Tatars are the most independent-oriented nation in Russia. Tatarstan also was not a signatory to the Federation Treaty, and did not participate in the 1993 elections and referendums. However, Tatar nationalists have all along recognised that the location of their homeland makes complete independence from Russia impossible, and have recently signed a bilateral treaty with Moscow that provides Tatarstan with relatively greater sovereignty but still within the Russian Federation (Teague, 1994; Raviot, 1994; Fayzulina, 1994).

Other nations in Russia whose homelands were autonomous republics have also become relatively more separatist, including Sakha (Yakutia), Tuvinia, Bashkortostan and Buryatia. However, all were signatories of the Federation Treaty, and most are pressing for a more confederal Russia rather than outright secession and independence. The titular élites in these republics have also renegotiated bilateral treaties with Moscow that provide them with greater sovereignty than they received in the Federation Treaty. Russia's willingness to renegotiate relations with the non-Russians in the republic, and the non-Russian élites' willingness to accept something less than complete independence from Russia, are indicators that Chechnya will not become a catalyst for the disintegration of the Russian Federation.

Conclusions

The history of the USSR was one of nationalisation, culminating in separatist nationalism and the disintegration of the state along national territorial lines. Since independence, nation-state building projects have become the predominant political action programmes in most of the NIS (i.e. state nationalism). This in turn has given rise to unification nationalism/irredentism, and to autonomist–separatist nationalism among non-titular nationalists who claim a certain part of the state as their ancestral homeland. The nationalism and ethnoterritorial conflicts that are present in the former USSR today are thus the products of an interactive process that developed first between centre and union republics – and between Russians and members of the titular nations – and is now developing between dominant and subordinate nations in the NIS. These conflicts are not the products of primordial animosities or ancient tribal hatreds, and they can be successfully managed. However, the trend line has been towards conflict escalation, as titular nationalists reject the legitimacy of territorial autonomy demands made by non-titular groups, and as the latter in turn have become more oriented towards irredentism and separatism.

Events in the former USSR also indicate that the relationship between socio-economic development and national identity that has been part of a core set of beliefs in both the West and in the USSR is highly dubious. The modernisation thesis that ethnonational identity would disappear with modernisation has clearly not occurred. Instead, ethnic group identity based more on a backward-looking sense of common origins has been converted into national identity, with its forward-looking sense of common destiny. In the process, ethnic identity has become politicised and territorialised in a new way, and national identity and nationalism have become dominant political interest groups and ideologies in the late twentieth century. Modernisation will not solve the national problem.

Similarly, equalisation will not solve the national problem. The precepts of internal colonialism, that national identity continued to have meaning in modernised states because of international inequality, and that equalisation would lead to the erasure of national identity, have not proven accurate in practice. International equalisation was occurring in the USSR, but this tended to coincide with a rising sense of national self-consciousness and growing international tensions, not with the internationalisation of the

population. National identity does not survive due to inequality, although inequality can serve to worsen international relations. Nationalists seek to gain a dominant or hegemonic status in their ancestral homelands, not an equal standing with national others. It may even be argued that equalisation tends to exacerbate international relations in a multi-homeland setting, because it brings a greater number of upwardly mobile indigenes into more intense contact and competition with élite representatives of non-titular groups. In the USSR, it was the most socio-economically developed nations, and the most socio-economically advanced segment of each nation's population (the intelligentsia) that were most nationalistic, and most intent on gaining independence/dominance within their homelands (Kaiser, 1994a). The lessons learned from the Soviet case indicate that we cannot look to socio-economic development and equalisation as means of solving national problems.

One solution recently proposed in Russia is to do away with the non-Russian republics (Petrov, 1994), and the new Russian constitution may be seen as a step in this direction. Advocates of this line are of the opinion that the federalisation of Russia and the USSR along ethnoterritorial lines is at the root of the new national problems which led to the disintegration of the USSR and is leading to similar disintegrative processes in Russia. However, once national identity exists and a sense of homeland develops, the forcible elimination of autonomy for nations in their homelands is certain to be counter-productive, and will accelerate the pace of disintegration in the state (Kaiser, 1991, 1994a). On the other hand, granting greater autonomy for nations in their homelands cannot be viewed as a solution to the national problem (as it was by Lenin). Federalism along national territorial lines should be viewed as one instrument in international conflict management, and if successfully utilised can lead to the amelioration of conflicts, even if it is not in and of itself a solution to them. With the exception of Chechnya, Moscow has adopted this strategy of flexible federalism in dealing with the autonomist nationalism of the indigenous non-Russians with republics in the state.

At present, titular nationalists continue to behave in a relatively exclusionary way towards the non-titular segment of their states' population, and the non-titular members have become more oriented towards opposition in response. Conflict escalation is likely to continue in the near future at least. However, in a few of the NIS, and particularly in Lithuania, early exclusionary nationalism has given way to more inclusive policies. The development of international relations in Lithuania suggests that once a titular nation has secured hegemony within its homeland (i.e. has gained control over its fate), the more exclusionary nationalistic policies can be compromised in favour of a more inclusive political strategy towards non-titular groups. Unfortunately, to date the titular political élites in most of the NIS are moving in the opposite direction, and it is difficult to be optimistic that peaceful, non-coercive international conflict management strategies will become commonplace any time soon.

References

Administratsii Prezidenta Rossiyskoy Federatsii (1993), *Konstitutsiya Rossiyskoy Federatsii*. Moscow: Yuridicheskaya Literatura.

Anderson, J. (1988), "Nationalist ideology and territory", in Johnston, R. *et al.* (eds), *Nationalism, Self-determination and Political Geography*, London: Croom Helm, pp. 18–39.

Arutyunyan, Yu. and Bromley, Yu. (eds) (1986), *Sotsial'no-kul'turnyy oblik sovetskikh natsiy.* Moscow: Nauka.

Barghoorn, F. (1956), *Soviet Russian Nationalism.* New York: Oxford University Press.

Bond, A. and Sagers, M. (1990), "Adoption of law on economic autonomy for the Baltic republics and the example of Estonia: a comment", *Soviet Geography*, 31: 1–10.

Breuilly, J. (1993), *Nationalism and the State*, 2nd edn. Manchester, UK: Manchester University Press.

Bromley, Yu. (ed.) (1977), *Sovremennyye etnicheskiye protsessy v SSSR.* Moscow: Nauka.

Bromley, Yu. (1983), "Etnograficheskoye izucheniye sovremennykh national'nykh protsessov v SSSR", *Sovetskaya Etnografiya*, No. 2: 4–14.

Bromley, Yu. (1984), *Theoretical Ethnography.* Moscow: Nauka.

Carrere d'Encausse, H. (1978), *The Decline of an Empire.* New York: Harper & Row.

Chinn, J. and Kaiser, R. (1996), *The Russians as a New Minority in the Newly Independent States of the Former USSR.* Boulder, Colo.: Westview Press.

Connor, W. (1972), "Nation-building or nation-destroying?", *World Politics*, 24: 319–55.

Connor, W. (1978), "A nation is a nation, is a state, is an ethnic group is a . . .", *Ethnic and Racial Studies*, 1: 377–400.

Connor, W. (1986), "The impact of homelands upon diasporas", in Sheffer, G. (ed.), *Modern Diasporas in International Politics*, London: Croom Helm, pp. 16–45.

Conquest, R. (1970), *The Nation Killers.* New York: Macmillan.

Conquest, R. (ed.) (1986), *The Last Empire.* Stanford, Calif.: Hoover Institution Press.

Crow, S. (1993), "Kozyrev on maintaining conquests", *RFE/RL News Briefs*, 2(41): 7.

Drobizheva, L. (1991), "Etnicheskoye samosoznaniye russkikh v sovremennykh usloviyakh: ideologiya i praktika', *Sovetskaya Etnografiya*, No. 1: 3–13.

Dzyuba, I. (1970), *Internationalism or Russification?*, 2nd edn. London: Weidenfeld & Nicolson.

Fayzulina, G. (1994), "Tatarstan: 'analogov ne imeet' ", *Moskovskiye Novosti*, No. 8: 10.

FBIS (1993), "Grachev arrives in Dushanbe, discusses border situation", *FBIS-SOV*, 19 July, No. 136: 63.

Foye, S. (1993), "Russian forces to defend against fundamentalism", *RFE/RL News Brief*, 2(8): 8.

Foye, S. (1994), "Yeltsin on near abroad, relations with west", *RFE/RL Daily Report*, No. 84.

Ginzburg, A., Ostapenko, L., Savoskul, S. and Subbotina, I. (1993), *Russkiye v novom zarubezh'e: Srednyaya Aziya.* Moscow: Tsentr po Izucheniyu Mezhnatsional'nykh Otnosheniy, Institut Etnologii i Antropologii.

Gleason, G. (1990a), *Federalism and Nationalism.* Boulder, Colo.: Westview Press.

Gleason, G. (1990b), "Marketization and migration: the politics of cotton in Central Asia", *Journal of Soviet Nationalities*, 1: 66–98.

Goble, P. (1990), "Gorbachev, secession, and the fate of reform", *RL Report on the USSR*, 2(17): 1–2.

Gorbachev, M. (1989), "O natsional'noy politike partii v sovremennykh usloviyakh", *Doklad i zaklyuchitel'noye slovo no plenume TsK KPSS 19, 20 Sentyabrya 1989 goda.* Moscow: Politizdat.

Gosudarstvennoye Statisticheskoye Agentstvo Kyrgyzskoy Respubliki (1993), *Demograficheskiy yezhegodnik Krygyzskoy Respubliki 1992.* Bishkek: Gosudarstvennoye Statisticheskoye Agentstvo Kyrgyzskoy Respubliki.

Grigulevich, I. and Kozlov, S. (eds) (1981), *Ethnocultural Processes and National Problems in the Modern World.* Moscow: Progress.

Guboglo, M. (1969), "O vliyaniii rasseleniya na yazykovyye protsessy", *Sovetskaya Etnografiya*, No. 5: 16–30.

Guboglo, M. (1984), *Sovremennyye etnoyazykovyye protsessy v SSSR.* Moscow: Nauka.

Gudkov, L. (1990), "Russians outside Russia", *Moscow News*, No. 41: 7.

Hennayake, S. (1992), "Interactive ethnonationalism: an alternative explanation of minority ethnonationalism", *Political Geography*, 11: 526–32.

Hodnett, G. (1967), "The debate over Soviet federalism", *Soviet Studies*, 18: 458–81.

Kaiser, R. (1991), "Nationalism: the challenge to Soviet federalism", in Bradshaw, M. (ed.), *The Soviet Union: a New Regional Geography?*, London: Belhaven, pp. 39–65.

Kaiser, R. (1994a), *The Geography of Nationalism in Russia and the USSR*. Princeton, NJ: Princeton University Press.

Kaiser, R. (1994b), "Ethnic demography and interstate relations in Central Asia", in Szporluk, R. (ed.), *The international politics of Eurasia. Volume 2: The influence of national identity*, Armonk, New York: M.E. Sharpe, pp. 230–65.

Kaiser, R. and Chinn, J. (1995), "Russian–Kazakh relations in Kazakhstan", *Post-Soviet Geography*, 36, 257–73.

King, C. (1994), "Moldovan identity and the politics of pan-Romanianism", *Slavic Review*, 53(2): 345–68.

Kipp, J. (1994), "The Zhirinovsky threat", *Foreign Affairs*, 73(3): 72–86.

Kolossov, V. (1992), *Ethno-territorial Conflicts and Boundaries in the Former Soviet Union*, territory briefing 2. Durham, UK: University of Durham IBRU Press.

McAuley, A. (1991), "Economic constraints on devolution: the Lithuanian case", in McAuley, A. (ed.), *Soviet Federalism: Nationalism and Economic Decentralisation*, Leicester and London: Leicester University Press, pp. 178–95.

Meyer, A. (1965), *The Soviet Political System*. New York: Random House.

Motyl, A. (1992), "From imperial decay to imperial collapse", in Rudolph, R. and Good, D. (eds), *Nationalism and Empire*, New York: St Martin's Press, pp. 15–44.

Orridge, A. and Williams, C. (1982), "Autonomist nationalism: a theoretical framework for spatial variations in its genesis and development", *Political Geography Quarterly*, 1: 19–39.

Petrov, N. (1994), "Ethno-territorial conflicts in the former USSR and possibilities for their regulation", paper delivered to the Kennan Institute for Advanced Russian Studies, Washington, DC.

Portugali, J. (1988), "Nationalism, social theory and the Israeli/Palestinian case", in Johnston, R. *et al.* (eds), *Nationalism, Self-determination and Political Geography*, London: Croom Helm, pp. 151–65.

Pravda (1990), "Zakon Soyuza Sovetskikh Sotsialisticheskikh Respublik o poryadke resheniya voprosov, svyazannykh s vykhodom soyuznoy respubiki iz SSSR", 7 April.

Raviot, J.-R. (1994), "Types of nationalism, society, and politics in Tatarstan", *Russian Politics and Law*, 32(2): 54–83.

Rostow, W. (1967), *The Dynamics of Soviet Society*. New York: W.W. Norton.

RUKH (1989), *RUKH Program and Charter* (Ellicott City, Md: Smoloskyp).

Salikov, R., Kopylov, I. and Yusupov, E. (1987), *Natsional'nyye protsessy v SSSR*. Moscow: Nauka.

Silver, B. (1986), "The ethnic and language dimensions in Russian and Soviet censuses", in Clem, R. (ed.), *Research Guide to the Russian and Soviet Censuses*, Ithaca: Cornell University Press, pp. 70–97.

Smith, Anthony (1986), *The Ethnic Origins of Nations*. Oxford: Basil Blackwell.

Smulders, M. (1990), *Who owes Whom? Mutual Economic Accounts between Latvia and the USSR, 1940–1990*. Riga: The Economic Reform Commission of the Council of Ministers of the Republic of Latvia.

Snezhkova, I. (1982), "K probleme izucheniya etnicheskogo samosoznaniya u detey i yunoshestva", *Sovetskaya Etnografiya*, No. 1: 80–8.

Solchanyk, R. (1992), "Ukraine, the (former) center, Russia, and 'Russia' ", *Studies in Comparative Communism*, 25: 31–45.

Solzhenitsyn, A. (1990), "Kak nam obustroit' Rossiyu", *Komsomol'skaya Pravda*, July (special issue).

Suny, R. (1992), "State, civil society, and ethnic cultural consolidation in the USSR – roots of the national question", in Lapidus, G. *et al.* (eds), *From Union to Commonwealth*, Cambridge: Cambridge University Press, pp. 22–44.

Taagepera, R. (1989), "Estonia's road to independence", *Problems of Communism*, 38(6): 11–26.

Teague, E. (1994), "Russia and Tatarstan sign power-sharing treaty", *RFE/RL Research Report*, 3(14): 19–27.

Toshchenko, Zh. (1994), "Potentsial'no opasnyye tochki: etnopoliticheskaya situatsiya v Rossii v 1993 godu", *Nezavisimaya Gazeta*, No. 39: 5.

Troynitsky, N. (ed.) (1905), *Obshchiy svod' po imperii rezul'tatov razrabotki dannykh' pervoy vseobshchey perepisi naseleniya*, 2 volumes. St Petersburg: Tsentral'nyy Statisticheskiy Komitet'.

Volkov, V. (1993), "What will Ukraine's policy be like?", *Moscow News*, No. 42: 5.

Williams, C. and Smith, A. (1983), "The national construction of social space", *Progress in Human Geography*, 7: 502–18.

Wishnevsky, J. (1994), "Yeltsin tacitly admits creating Chechen opposition", *RFE/RL Daily Report*, No. 153.

Yemel'yanenko, V. (1994), "Poslednyaya voyna?", *Moskovskiye Novosti*, No. 8: 4.

3

Geopolitics, history and Russian national identity

Denis J.B. Shaw
University of Birmingham, UK

A glance at the map of the former USSR will show at once how well placed Russia is to dominate the other post-Soviet states. The geographical reasons are straightforward: Russia contains over three-quarters of the territory and over half of the population of the former USSR. But the reasons are also historical. Numerically, politically and culturally Russians dominated both the Soviet Union and the pre-1917 Russian Empire. In fact, in both states Russians constituted the "imperial" people. This has important implications for the role many Russians see their country playing now that the USSR is no more. The argument of this chapter is that the Russian Federation is a special case among post-Soviet states. Like the other post-Soviet republics, Russia needs to adjust to its geopolitical situation, but unlike them its imperial past and history of domination seem set to make that adjustment particularly difficult. The crux of the problem lies in how far Russians can agree that the presently constituted Russian Federation is a legitimate expression of Russian nationhood. Before enquiring into that issue, however, something must be said about Russia's origins and multiethnic character since these factors help explain many of the problems which face the country today.

Muscovite Russia: the origins of the Russian state

Oddly enough, Russia is both one of the newest and one of the oldest of Europe's states. As an independent state, the Russian Federation dates back only to 1991, though as a political formation it derives from decisions made by the Bolsheviks after they had seized power in the revolution of 1917. The Russian state as a state, however, has a much longer history. Essentially, the modern Russian state arose from the principality of Muscovy which was constructed around the city of Moscow in late medieval times and then gradually expanded to become an empire. For much of its history, the Russian state has been ruled from Moscow. However, from the early eighteenth century until 1918 (during most of which period it was known as the Russian Empire) Russia was ruled from St Petersburg on the Baltic.

One of the problems which has always plagued Russians is to decide exactly what it means to be "Russian". For example, before the rise of Muscovy, the Russian people and their close relatives, the Ukrainians and Belarusians (who are all Eastern Slavs) were divided up among a series of principalities, and from the ninth to about the twelfth centuries AD these principalities together formed a loosely structured state known as Kievan Rus. Ukrainian nationalists have long argued that Rus was essentially Ukrainian and that the Russians formed a separate and peripheral element in the Kievan state. In this way Ukrainians have hoped to deny any historical justification for Russian claims to Ukraine. However, both modern Russian nationalists and past rulers of Muscovy have seen things differently. By the time of Muscovy's rise in the late medieval period, the western part of what had been Rus (corresponding to modern Ukraine and Belarus) had fallen under the sway of the Lithuanians and then the Poles. Muscovite rulers began to advance claims to this western territory, arguing that the region was historically part of "Russia" and had been usurped by foreign powers. Later, in the seventeenth and eighteenth centuries, the Russian tsars succeeded in annexing what is now Ukraine and Belarus and asserted that this was merely a "recovery" of historic Russian land. Similar attitudes lie behind the arguments of some present-day Russian nationalists who resent the post-1991 independence of Ukraine and Belarus and who feel that they should be absorbed into a greater Russia. Needless to say, such ideas are fiercely contested by those, both inside and outside Russia, who are fearful of a revival of Russian imperialism.

What the rulers of Muscovy asserted and eventually achieved was their *de facto* right to rule over all "Russians". Their uncompromising attitude was no doubt forged during the two centuries which followed the catastrophic Mongol invasion of AD 1237–42. During these centuries Russia (i.e. the eastern part of what had been

Geography and Transition in the Post-Soviet Republics. Edited by M.J. Bradshaw.
© 1997 M.J. Bradshaw & Contributors. Published 1997 by John Wiley & Sons Ltd.

Rus) formed part of a Mongol–Tatar empire, known as the Golden Horde, and the various Russian princes were obliged to pay homage to the khan or ruler of this empire. However, the princes or rulers of the small principality of Moscow proved particularly adept at gaining the khan's favour and cajoling or browbeating the other Russian rulers into submission. Eventually Moscow's princes proved powerful enough to repudiate their dependence on the Golden Horde. Some modern historians have argued that the princes of Muscovy had in effect replaced the khan as rulers of Russia, assuming in the process some of his "Eastern" characteristics (Pipes, 1977, pp. 73–84; Vernadsky, 1969). Thus they ascribe the well-known autocratic and authoritarian tendencies of later Russian rulers to their Mongol–Tatar forebears. According to this view, if Russia had been unified not by Moscow but, say, by Novgorod, the wealthy and westward-oriented trading city not far from the shores of the Baltic, then Russia would have become far more "Western" and European than was in fact the case. In the event, Novgorod's independence was crushed by the Muscovite prince Ivan the Great in 1478.

A further factor which no doubt helped to give Moscow's rulers imperial aspirations was the religious one. Christianity first came to Rus in the year AD 988 in its Eastern Orthodox form with the baptism of Prince Vladimir of Kiev and the "conversion" of his subjects. From 1326 the Metropolitan or head of the Russian Orthodox Church resided in Moscow, thus helping to support the political aspirations of that city's rulers. As a frontier state, in frequent conflict with the Tatars (who were mainly Muslim), Muscovy easily succumbed to religious zealotry. Such tendencies were further encouraged by international events. In 1439, at the Church Council of Florence, the Eastern Orthodox Church was united with the Roman Catholic one, an event which horrified Muscovite traditionalists. Only a few years later, in 1453, Constantinople, the spiritual centre of Eastern Orthodoxy, was conquered by the Muslim Turks. Devout Russians saw this as divine punishment for the heresy perpetrated by the Council of Florence. Eventually, this gave rise to official adoption by Russia's rulers of the doctrine of the Third Rome. According to this doctrine, the spiritual centre of true Christianity had originally moved from Rome in Italy, which had succumbed to the heresy of Roman Catholicism, to Constantinople, the capital of the Eastern Roman or Byzantine Empire. However, Constantinople in turn had succumbed to heresy and been captured by the Turks. Thus the spiritual centre had now moved to Moscow, capital of the only remaining independent Orthodox realm, thus becoming the "Third Rome". As a symbol of this fact, in 1547, Ivan the Terrible of Moscow was crowned as tsar, which is the Russian word for Caesar (Billington, 1970, pp. 49–77; Milyukov, 1974, pp. 21–7).

It is difficult to know how far the Third Rome doctrine had any practical effects in terms of the tsars' policies of territorial expansion (Obolensky, 1950, pp. 46–7). What it does seem to have done, together with other ideas derived from Orthodoxy, is to encourage a sense of superiority relative to the outside world and an intolerance of foreign ways. When in time Russians found themselves challenged by new ideas from abroad, as reflected by the growing power of the western European states, they found great difficulty in reconciling such new ideas with Russian traditions. In other words, they began to experience a crisis of identity. Something else which eventually tended to produce the same effect was Russia's transformation into a multi-ethnic state. And it was Ivan the Terrible, who had done so much to unite Russians around the banner of Orthodoxy, who initiated this transformation. In 1552 Ivan's armies invested and then seized the Tatar state of Kazan on the Volga. From this time onwards Russians found themselves ruling over increasing numbers of alien peoples as their territories continued to expand. It is this aspect of Russia's developing character which must now be examined.

Russia as a multi-ethnic state

By Ivan the Terrible's reign (1533–84) Muscovy had already expanded greatly beyond the few thousand square kilometres it had occupied in the fourteenth century, having absorbed all the other Russian principalities. By this stage it had also occupied Novgorod's extensive territories in the coniferous forests of the northern part of European Russia, which Novgorod had exploited for their fur-hunting. After the fall of Kazan in 1552, Muscovy continued to expand geographically in virtually every direction well beyond the bounds of the traditional Russian homeland, a process which took several centuries to complete. Thus the Russia-state moved north-eastwards and eastwards across Siberia and the Far East, eventually reaching Alaska in North America; westwards into what is now Belarus, Ukraine, the Baltic provinces (the future Baltic states) and even into Poland and Finland; southwards to the shores of the Black Sea and into the Transcaucasus; and south-eastwards into Central Asia. Over the course of about 300 years, therefore, Moscow went on to build one of the world's great empires. Partly as a result of this, the proportion of Russians in the population ruled by the tsars declined considerably: from about 70% of the total in 1719 to a mere 44.6% in 1917. At the same time, the total number of the tsars' subjects increased manyfold: from some 14 million to about 170 million over the same period.

As was the case in other empires founded by the European states over a similar period, there was considerable migration and settlement by the "imperial" people – namely the Russians, who were also joined by numerous Ukrainians, Belarusians and some other groups – well beyond the bounds of the original homeland. Thus very large numbers of Russians and Ukrainians settled to the south in the fertile European steppe, and also to the east in the southern part of Siberia, converting the latter in particular into an extension of the European Russian homeland. In both these cases, the pre-existing native peoples were largely absorbed, removed or greatly reduced in numbers. Elsewhere, however, Russian settlement was less significant. Sometimes this was because of environmental disadvantages such as in the Far North, where relatively few Russians were attracted before 1917, or in the arid regions of Central Asia. In other areas the opportunities available to Russian settlers were limited by the fact that the regions were already well populated by non-Russians. This was the case in the Transcaucasus, in much of Central Asia, and in parts of the west (western Ukraine and Belarus, the Baltic

provinces, Poland and Finland). In many parts of the Transcaucasus and Central Asia, Russian settlement was restricted to towns where industry, administration or the military offered employment opportunities.

The tsars were prepared to make few concessions to any national aspirations or to the cultural peculiarities of their non-Russian subjects (Raeff, 1971). As far as the tsars were concerned, Russian values and ways of life served as the norm to which all subjects in the empire were expected to conform sooner or later. The Orthodox faith, for example, was considered to be the only true one, providing the justification for the tsar's powers and for Russian rule over the other peoples. Non-Russian peoples and their ways were grudgingly tolerated rather than granted equal status with the Russians and their customs; occasionally, non-Russians were persecuted. Only in the western territories, in Poland, Finland and the Baltic provinces, were the tsars prepared to grant a measure of autonomy and special recognition of the rights of their non-Russian subjects in deference to the latters' strong Western heritages. (A more benign interpretation of tsarist nationalities policy will be found in Zubov (1994).) But such concessions were regarded as temporary and were later revoked in response to rising national movements among these peoples. At the very end of the tsarist period further concessions had to be granted in the wake of revolutionary unrest around 1905, but again the authorities refused to consider such concessions permanent.

When Lenin and the Bolsheviks seized power in Russia in 1917, they decided to grant some recognition to the national rights and aspirations of non-Russian peoples. This was not because the Bolsheviks believed in nationalism – indeed, as working-class internationalists they were very much against it – but because they thought that this was the best way of attracting the support of minority peoples for the Russian Revolution and against its many enemies both at home and abroad. Eventually, their new state (the USSR) which replaced the old Russian Empire was organised as a kind of federation of nominally equal workers' republics. However, the degree of autonomy allowed to these republics was extremely limited in practice; all were firmly in the grip of

Lenin and the communists, based in Moscow. Even less autonomy was allowed to the autonomous republics (ASSRs), regions and districts which were set up within the larger (union) republics to represent the national aspirations of smaller minorities.

By far the biggest and most populous of the new union republics which came into being after 1917 was the Russian Federation or RSFSR. This included not only the original Russian homeland to the west of the Urals but also territories beyond, such as a good part of the European steppe and southern Siberia where there had been dense Russian settlement subsequently. Moreover, the new republic contained many areas, such as in the Far North, the middle Volga and the North Caucasus, where Russians were either in a minority or where their numbers barely exceeded those of the native peoples. These areas were included in the new Russia either for geographical reasons (for example, they were surrounded by Russian territory) or because the native peoples were not considered to be numerous enough to merit their own union republics. It was in such areas that autonomous republics, regions and districts were organised within the Russian Federation. Thus the Russian Federation, which was part of the Soviet Federation, the USSR, was itself a federation containing many non-Russian peoples as well as Russians. This has given rise to numerous problems, as noted later in this chapter.

Unfortunately, when the boundaries of the Russian Federation were being demarcated in the years after the 1917 revolution, they were drawn so as to leave many Russians living beyond them in other union republics. Numerous Russians, for example, lived in northern Kazakhstan and in eastern Ukraine, partly as a result of industrial and agricultural migrations in the later tsarist period. This situation was exacerbated after the first five-year plan was launched in the late 1920s. Thus many Russians migrated into Ukraine, the western republics (including Estonia and Latvia, annexed by Stalin in 1940 after they had enjoyed a period of independence from 1917) and elsewhere to take up employment in the new industries. Others migrated into northern Kazakhstan as part of the Virgin Lands agricultural settlement scheme in the 1950s. All this mattered

little as long as the USSR held together since, wherever they lived in the USSR, Russians were still ruled from Moscow and enjoyed its protection. With the breakup of the USSR in 1991, however, such people have suddenly become minorities in the non-Russian republics (Harris, 1993). This has important implications not only for the Russian minorities concerned but also for the policy of the Russian Federation.

The Russian Federation and Russian identity

We have made the point that Russians seem to have found it difficult throughout much of their history to define what it means to be "Russian". This may have something to do with their geographical situation and lack of clearly defined frontiers. From the time that the Eastern Slavs migrated on to the East European plain in the first millennium AD, Russian settlement flowed around and intermingled with pre-existing settled peoples, often with few natural or even human barriers to prevent it. The cultural or ethnic differences which may or may not have existed between the Eastern Slav peoples in the early period also seemed unimportant and this fact has since been used by some to argue for a "natural" unity between Russians, Ukrainians and Belarusians, as we have seen. Only when Russians came up against well-organised resistance by other peoples, as in the case of the Tatar and other nomads in the steppelands to the south, did they become acutely aware of cultural differences. Even here, however, there was generally little incentive to make special provision or allowance for such differences.

Geography thus helped to foster an important difference between the way Russians regarded their empire and the way other Europeans regarded theirs. In the latter case, the ocean provided an obvious point of demarcation between the homeland and the overseas empire and with some exceptions (such as the British in Ireland or the French in Algeria), western Europeans had relatively little difficulty in deciding where their homelands ended, and where their imperial possessions began. Clearly for the Russians the situation is much less straightforward because of the land-based nature of their erstwhile empire (Zubov, 1994, pp. 37–44). Some present-day commentators have even suggested that, because of confusion over where their homeland ends and that of other peoples' begins, the Russians cannot help but be imperialists. This seems to the present author to push the argument too far, an example of anti-Russian prejudice. But if we think of the difficulties western Europeans have had in agreeing their frontiers (and still have, in the case of Northern Ireland, for example), then it should come as no surprise to learn that the Russians and their neighbours have similar problems.

It is because of this particular imperial history that many Russians, particularly those of nationalist or communist inclination, have questioned the legitimacy of the Russian Federation (Walker, 1992; Shaw, 1993). The Russian Federation, it is pointed out, came into being as a result of decisions made by the Bolsheviks in their early years of power, and anti-communists in particular are apt to question why decisions made by a now discredited communist regime should be considered sacrosanct. One frequently cited criticism of the present-day boundaries of the federation cites the existence of up to 25 million ethnic Russians living in other republics. In some instances, such as in north-eastern Estonia, Crimea in Ukraine, and in some parts of northern Kazakhstan, the Russians form a compact population with a local majority adjacent to the Russian Federation border – clearly a case, in the opinion of some, for frontier adjustment. Another criticism alludes to the "artificial" way in which Russians have now been divorced from their close kinsfolk, the Ukrainians and Belarusians with whom, as we have seen, there are many historic links (though by no means all friendly ones) (see, for example, Solzhenitsyn, 1991, pp. 19–21). Finally, many Russian nationalists and communists have argued that the 1991 breakup of the USSR was without legal foundation and have called for its restoration, seeing this as the best hope for the future prosperity and security of all the peoples of the former USSR, but perhaps particularly of the Russians.

Just as there are people who would wish to see the Russian Federation expanded to take in all ethnic Russians, or dissolved into some imperial superstate, so there are those who would wish to see it diminished. The latter would obviously include some members of the Russian Federation's minorities. At the 1989 census, Russians constituted 81.5% of the federation's total population, leaving some 18.5% or 27 million people divided among more than 90 nationalities. Over half of the federation's territory is occupied by various kinds of "autonomous" territory – namely, 21 republics, 1 autonomous region and 10 autonomous districts. There have been many arguments about how much autonomy these subsidiary units are entitled to, and by no means all of these have been solved by the Federation Treaty, signed by 18 out of (then) 20 republics in March 1992, or by the new Constitution, adopted towards the end of 1993. The most extreme case of attempted secession is that of the North Caucasian republic of Chechnya which has been the victim of invasion and a bloody war. Since the Russian Federation is the most obvious successor state to the old USSR, it is no wonder that some minority groups regard it as an equally unacceptable form of imperial power and view attempts to keep it united as efforts to prop up a decaying empire.

Factors like its own uncertain history, its somewhat unstable democracy, a lack of experience of federal and constitutional forms, and disagreement over the correct relationship between ethnic and individual rights, may certainly serve to threaten the cohesion of the federation (Shaw, 1993). Doubts about the proper form and purpose of the federation are reflected in its twin names: according to the Constitution, the state is called both "Russia" and the "Russian Federation". What lies behind this are disagreements over whether the state is meant to represent primarily the ethnic Russian people, or all its citizens, no matter what ethnic group they happen to belong to. Until this matter is resolved, there must be considerable potential for future instability.

Russia and the wider world

The new Russia must inevitably try to find some means of cohering as a state and building stable relations with other countries. The latter means, in the first instance, its immediate, post-Soviet neighbours in what is sometimes referred to as the "near abroad". Some of the influences which might threaten to draw Russia into conflict with its near neighbours have been described already. First, there is the problem of the 25 million Russians living there (rather fewer now as a result of migrations back to Russia). Fears that such people might be the victims of discrimination, of policies of forced assimilation, or even of "ethnic cleansing" have encouraged some of them to appeal to the Russian government for protection. Needless to say, such appeals find a ready hearing among nationalist groups in the Russian Federation, and there have been a number of difficult incidents in which Russian and neighbouring governments have disagreed about the rights of the Russian minorities. For their part, some of the other post-Soviet states have appeared to regard their Russian minorities as a potential "fifth column" and feared that they might furnish an excuse for interference by Moscow in their internal affairs. Often enough, what also encourages such disputes is disagreement between Russia and its neighbours over other matters, such as strategic and military ones in the case of Crimea in Ukraine, or transit rights in that of the Baltic states.

Further reasons for discord between Russia and its near neighbours have included disputes about frontiers and territories, disagreements about ex-Soviet property, military and security matters, environmental issues, trade-related problems, and others. The breakup of the USSR was attended by many difficulties, particularly because of the fragmentation of what had previously been a unified economic space (Bradshaw, 1993). Most of the successor states therefore agreed to form a loose association, known as the Commonwealth of Independent States, designed to help overcome such problems and smooth the post-Soviet transition. However, the CIS has remained a rather shadowy entity, partly because of fears by the non-Russian states that it might be used by Russia to dominate them. As it is, the Russian Federation, with its enormous resource potential and formidable military might, has been able to exert considerable influence over many of the other states which are

often dependent upon it for energy and other needs. Russia also has security concerns, particularly in regard to the serious disturbances and civil wars which have afflicted several of the states lying to its south. There are fears that the disorders among some of the Islamic peoples of these regions might spread among the Muslims living in Russia itself. This is probably one of the reasons for the hard line pursued in Chechnya. Russian forces have been influential in seeking to sort out disputes in the Transcaucasus, Tajikistan and Moldova, and Russian leaders have sought international approval for their security role in the post-Soviet arena. But, perhaps inevitably given Russia's history, such ventures have stimulated fears that Russia might be pursuing its old imperial agenda once again.

Whereas the old USSR regarded itself as a rival of the West and sought to gain world influence at the expense of the Western capitalist powers, the new Russia has been rather more accommodating to Western interests, opening itself up to world trade and seeking Western help to restructure its economy. Russia remains, however, a nuclear power, albeit a gradually disarming one, and the old suspicions about its ambitions and pretensions are not easily laid to rest. Moreover, Russia inevitably has its own national interests to pursue, and these may not always accord with those of the United States and its allies (just as the national interests of those countries do not always accord with one another). Russia's Muslim population, and the country's geographical proximity to Islamic countries lying to its south and in the Middle East more generally, mean that its interests in these regions are almost bound to be different from those of the USA. Russia has historic ties with Serbia and other Orthodox countries in the Balkans, causing it to view the breakup of Yugoslavia in a rather different light. And historic fears of isolation, and concerns about the (by no means entirely unjustified) hostility of some of the populations of Central and Eastern Europe have naturally led to worries about the implications of plans to admit some of the ex-communist states in the latter region to full membership of NATO. Of course future Russian policy very much depends upon who rules in Moscow. But, whatever happens, Russia somehow has to learn to live with the consequences of its own history. At the same time, it would be quite unrealistic to expect a country of the enormous size and diversity of Russia to have geopolitical interests which are entirely the same as those of the West.

Russia and the West: the historical dimension

From a historical perspective, trying to find the right kind of relationships between Russia and the rest of the world is part of the age-old search for Russian identity and a way of life acceptable to its people. In a previous section it was noted that, in unifying Russia, the rulers of Muscovy made use of the Orthodox faith in boosting their own legitimacy and political aspirations. By the seventeenth century, however, the apparent unity of the Russian realm was already being challenged by the import of ideas from abroad, particularly from the West. In the middle of that century the Russian Church was rent by schism. By the century's end, Russia was being subjected to policies of modernisation by Tsar Peter the Great who was eager to strengthen his country militarily and to transform it into a major European state. In this he was only partially successful.

Peter the Great was the first, but by no means the last, ruler of Russia to confront what was to become its historic problem, that of relative backwardness. To Russia's west in Peter's day lay a series of European states whose power was clearly on the increase and which might in time become a serious threat. Meanwhile Russia itself, after centuries of virtual isolation, seemed to wallow in self-satisfied slumber. To Peter the only possible answer appeared to lie in forced modernisation. But while this did have the effect of boosting Russian power and laying the foundations for future progress, the changes failed to become self-sustaining and after Peter's death other European states continued to forge ahead.

In thus trying to transform his realm, Peter left a legacy whose effects can still be detected today. One aspect of this was a serious social dichotomy between a small, educated and Westernised élite and the mass of the Russian people, most of whom were either indifferent or hostile to pol-

icies of reform. Even today Russian intellectuals often express doubts about the extent to which the masses are converted to reform. Another effect was to sow the seeds of doubt about the future thrust of Russian development. These seeds began to sprout during Peter's reign when his reforms were opposed by reactionary elements and the tsar felt constrained to have his own son, a religious conservative, put to death. Later on in the nineteenth century Peter's personality and reputation became a matter of sharp dispute between those who regarded his reforms as inherently progressive and essential to Russian well-being (known as Westernisers) and those who felt that they had done immense harm to Russian unity and its integral civilisation (the Slavophils). By the early years of the twentieth century the arguments had become more complicated, but in essence they still concerned how far it was desirable and possible for Russia to follow the Western path of progress (Walicki, 1979).

Discussions concerning Russia's future were cut off by the 1917 revolution. The Bolsheviks, who were Marxists, proceeded to apply Marxist policies to Russia, thus following a creed which had a Western European rather than a Russian provenance. However, under the influence of the Bolshevik leaders Lenin and Stalin, and in response to the practical necessities of having to rule in Russia, what finally emerged was very much a Russianised version of Marxism. Stalin in particular, facing international isolation in the midst of hostile capitalist powers, talked of the necessity of building "socialism in one country". This helped foster an idea which in fact has a long history in Russian thinking – the idea that Russia is somehow special and a source of enlightenment to its neighbours (Orthodox enlightenment in the Muscovite era, socialist enlightenment in the days of Stalin). And as in the Muscovite period, under Stalin and his successors this attitude helped foster a suspicion of the outside world and a reluctance to welcome foreigners and their dubious ways.

As long as Stalin was alive, manifestations of dissent from the official ideology were met with persecution and state terrorism. After his death in 1953 there was a re-examination of his heritage and criticism of at least some of the excesses which had characterised his method of rule. By the 1970s it was becoming apparent that the Soviet system could no longer guarantee the rises in living standards and the international political successes which had long been taken for granted. Thus dissent gradually became more widespread (Hosking, 1992, pp. 402–45). One very important factor which lay behind this was the growing contacts with the outside world, as the Soviet Union felt increasingly constrained to participate in the international market-place and as Russians became more exposed to foreign ways through the media and tourism (even though both remained strictly controlled by the authorities). Dissent took many forms, but what is especially interesting is the way in which some of the older arguments about Russia's future began to resurface. Thus some dissenters argued that the Soviet Union's best hope was to transform itself into a Western-type democracy. Others doubted the wisdom or possibility of this idea and felt that a reformed Soviet Union or Russia should seek its own development path, one that would be truer to what were seen as Russian traditions. Again, with the eventual rise of various non-Russian nationalisms in the periphery, Soviet leaders (most of whom were Russian) were no longer able to assume that Moscow's right to dominate this multi-ethnic society would continue to be accepted unquestioningly by the ethnic minorities.

The breakup of the Soviet Union in 1991 has challenged Russians once again to re-examine what it means to be "Russian", including their relationships with their neighbours and the wider world. While the demise of the old Soviet system is regretted by relatively few (at least among the younger generation), there is little agreement on what should come next. Thus Boris Yeltsin and his associates in the Russian government embarked on a programme of radical Westernisation as soon as the Soviet Union collapsed. Their aim was to transform Russia into a market-type economy and Western-type democracy as soon as possible. Like earlier Russian rulers, however, Yeltsin found the task far from easy and has since had to compromise with more cautious and nationalist elements. Many others regard the whole programme of marketisation and the attempt to circumscribe Russia's foreign policy ambitions with deep suspicion. There seems little doubt that the old state-dominated, communist-type system can never be revived. How far the new one which replaces it accords with common ideas of pluralist capitalism remains an open question.

Conclusions

This chapter has argued that Russia, like the other post-Soviet states, is having to adjust to a new world order but, as a former imperial power, is finding that adjustment particularly problematic. It has also suggested a number of historical reasons for that difficulty.

First, there is what we might call Russia's Muscovite–Orthodox heritage. The Russian state came into being in circumstances which largely isolated it from the outside world, especially from Europe. The religious heritage encouraged a sense that Russia was somehow special, which raised problems about how the country was to relate to the outside world once it became necessary to do so.

Second, there is Russia's heritage as a land-based and multi-ethnic empire, raising problems about what it means to be "Russian", about the rights of non-Russians within the Russian state, about how to ensure the cohesiveness of the state, and about the frontiers of the state, or where Russia should end and other countries begin.

Third, there is what we might call the heritage of Peter the Great, the question of Russian attitudes to Westernisation and modernisation, and the problems of social dichotomy which attempts to modernise Russia have so often brought in their wake.

Perhaps the most basic problem of all comes from trying to apply a simplistic model of "modernisation" to Russia as well as to other countries. That the world is in fact an enormously complex and varied place, and that simple solutions to development problems are almost always unsatisfactory, seems to be a lesson which politicians find difficult if not impossible to learn. Perhaps this is because it implies that their task is not an easy one and that they are likely to fail. Which makes it all the more important that geographers continue teaching it.

References

Billington, J.H. (1970), *The Icon and the Axe*. New York: Vintage Books.

Bradshaw, M.J. (1993), *The Economic Effects of Soviet Dissolution*. London: RIIA.

Harris, C.D. (1993), "The new Russian minorities: a statistical overview", *Post-Soviet Geography*, **34**(1), January: 1–27.

Hosking, G. (1992), *A History of the Soviet Union*, final edition. London: Fontana.

Milyukov, P. (1974), *Outlines of Russian Culture*, Vol. 3 *The Origins of Ideology*. Gulf Breeze, Fla: Academy International.

Obolensky, D. (1950), "Russia's Byzantine heritage", *Oxford Slavonic Papers*, 1: 37–63.

Pipes, R. (1977), *Russia under the Old Regime*. Harmondsworth: Peregrine Books.

Raeff, M. (1971), "Patterns of Russian imperial policy toward the nationalities", in Allworth, E. (ed.), *Soviet Nationality Problems*, New York: Columbia University Press, pp. 201–27.

Shaw, D.J.B. (1993), "Geographic and historic observations on the future of a federal Russia", *Post-Soviet Geography*, **34**(8), October: 530–40.

Solzhenitsyn, A. (1991), *Rebuilding Russia*. London: Harvill.

Vernadsky, G. (1969), *The Tsardom of Moscow, 1547–1682*. New Haven: Yale University Press.

Walicki, A. (1979), *A History of Russian Thought from the Enlightenment to Marxism*. Stanford: Stanford University Press.

Walker, E.W. (1992), "The new Russian Constitution and the future of the Russian Federation', *The Harrison Institute Forum*, **5**: 10 June.

Zubov, A.B. (1994), "The Soviet Union: from an empire into nothing?", *Russian Social Science Review*, **32**: 6–36.

References

4

Between space and race: rediscovering Russian cultural geography

Alexei Novikov
Moscow State University, Russia

Before one can talk about a cultural geography of Russia it is necessary to determine whether such a thing is possible within a Russian cultural context. Whether there is a subject for cultural geography – the differentiation of Russian culture across space – or whether it is just a methodological approach within geography oriented towards the search for spatial distinctions even when a society is not aware of its own territorial structure. Does Russian society realise its differentiation in space, and if it is so, how far and in what way?[1] To obtain an answer to this question in the context of Russia is very difficult. The rapid dynamics of social transformation, the absence of relatively stable and mature public opinion about crucial problems of the country's development, the instability of political institutions and preferences, a chaotic electoral geography, all of these factors make it very difficult to determine the level of spatial organisation of society, or its regional self-consciousness.[2]

Nevertheless, at the same time, taking into account historical, literary and philosophical sources, as well as the modern media, one can prove that in relation to their own country Russians have created stable mythologies. Most of them can be reduced to one of two points of view. From one side Russia is represented as a "monotonous faceless" space, from the other, just the opposite, as a "multi-ethnic and diversified" space (see Figure 4.1). In spite of the fact that these mythologies have all the attributes of social prejudices, and may even be treated as forms of "national delusion", these concepts are examples of the active social and cultural construction of space in Russia. Such attitudes provide an opportunity to treat Russian cultural geography not only as an academic discipline – as a collage of historical, economic and geographical facts – but as something which is produced by society itself.

Cultural monotony and facelessness

The notorious image of the Russian cultural landscape as a faceless place has become almost ubiquitous. In a recently published reader for geography students, called "Spaces of Russia" (Zamyatin and Zamyatin, 1994), the readings chosen suggest that the approach to Russia as a monotonous space can be read as cultural stereotype independent of the individual author's characteristics, such as their social position, place of birth and residence, etc. Fyodor Dostoevsky, a magnificent connoisseur of Russian national character, specified the absence of "sense of a place" in Russia. The French writer A. de Coustin described Russia as "a country without landscapes". The Russian Tsar Nikolas the First considered the great distance and monotonous spaces as a "disaster for Russia". In a similar vein, more recently, the post-modern Russian writer A. Bitov experienced a nightmare during his trip to Siberia (1993, p. 50):

Once I was travelling through the Western Siberian Lowlands. I woke up and glanced out the window – sparse woods, a swamp, level terrain. A cow standing knee-deep in the swamp and chewing, levelly moving her jaw. I fell asleep, woke up – sparse woods, a swamp, a cow chewing, knee-deep. I woke up the second day – a swamp, a cow. This was not space – it was a nightmare.

Even supporters of the "Eurasian" concept, who insist on a determining role for Russia's intermediate geographical location between Europe and Asia, are not willing to notice any spatial diversity in Russian culture. This can be confirmed by the words of L. Gumilyev, one of the foremost supporters of the "Eurasian" concept, as: "The border between Europe and Asia crosses the heart of every Russian." Thus, L. Gumilyev transforms the spatial construction of Russia into a psychological one; such universality denies the possibility of a spatially differentiated Russian culture. Within the Russian geographical community similar views are displayed in a more latent form. Despite enormous interest on the part of Russian geographers in the regional concept, and the regionalisation of Russia, almost none of them have thought to divide Russia on a cultural basis. This is not only because of a lack of desire, but, first of all, because of the lack of stimulating motives. The majority of works

Geography and Transition in the Post-Soviet Republics. Edited by M.J. Bradshaw.
© 1997 M.J. Bradshaw & Contributors. Published 1997 by John Wiley & Sons Ltd.

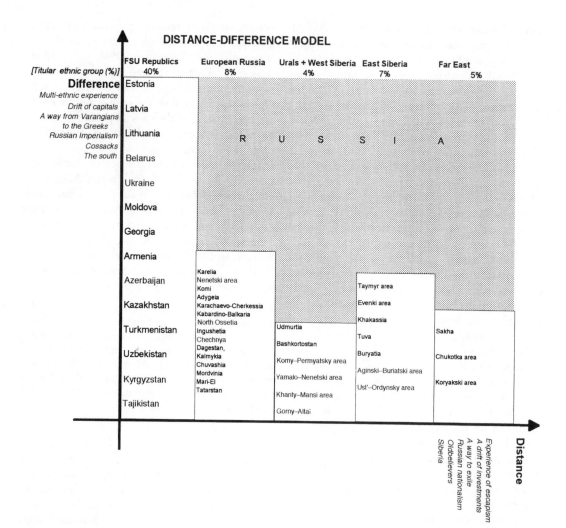

Figure 4.1 *Distance-difference model*

whose titles include "cultural geography" are concerned with the cultural infrastructure of Russia rather than with cultural images (Burg and Cukhchi, 1994).

Modern regional geography (*stranovedenie*) stands very close to cultural regionalisation of the territory, particularly the project, "Geography of the Fatherland" led by S. Rogachev. This has been organised like a colourful collage, made of historical, geographical, literary and visual material. However, the images of Russia and her regions which are being created in the "Geography of the Fatherland" are, to a large extent, the results of S. Rogachev's individual personality, rather than a vernacular understanding of what Russia is about produced by Russian society itself. The aim of Rogachev's project is to cultivate a sense of place among Russians, rather than to describe various parts of the country.

The phenomenon of the monotony of Russian cultural space was first clearly recognised and described by L. Smirnyagin (1995). According to his concept of the "aspatiality of Russian culture", cultural differences in Russia exist mostly at macro and micro levels. For example, a cultural difference between European Russia and Siberia is quite obvious. At the same time, the differences in the style of housing construction, in rituals and dialects are noticeable among small settlements (hamlets, villages, towns) across the whole of Russia. Nevertheless, distinctive regional communities of the middle size (at the meso level), comparable, say, to the American Deep South, Old South, or New England, seem absent in Russia. Political regions (*oblasts, krays* and autonomous republics) take the place of vernacular regional communities at the meso level. The administrative and political division of the territory appears to be the major framework for the manifestation of social and cultural differences at the meso level. In other words, any voluntary organisation of people into regional communities at this level is repressed by the spatial organisation of the state (centralised in the nearest past and increasingly federalist at present). The lack of meso-scale identity is one of the major factors promoting the phenomenon of Russia's disintegration because there is no good counterbalance to the regional disintegration of the state on the part of cultural and social regionalism. The latter being created by the voluntary manifestations of people may actually transcend and bind the administrative divisions of the state. Thus, the *oblast*-level system of boundaries may not have corresponded to the cultural landscape upon which they were imposed. But the contemporary reality is that many people now find themselves repressed by the regional state rather than by the central government. The essence of the transformation which Russia is undergoing at the moment could be symbolically represented as a struggle between the Russian people (*Russkii Narod* as an ethnic group) and Russian society (people who live within Russia). The former are being taken over and strongly supported by Russian spiritual culture, which unlike social culture plays a consolidating role. Russian literature and colloquial Russian language have achieved more than others in the unification of Russian culture; in people's minds language and literature play a crucial role in creating a national identity. According to M. Yampol'skiy (1993), literature and language in Russia are becoming a major national theme, or even myth, which holds the whole nation together. In public consciousness the idea of the Russian nation is being formed not by Yaroslav the Wise (who, incidentally, ruled in Ukraine), but by A. Pushkin, L. Tolstoy and F. Dostoevsky.

There is almost a mystique relationship between the size of Russia and the size of the most important Russian novel *War and Peace*. It is almost as if Russian writers should produce something comparable in size to Russia herself. After Tolstoy the role of the modern classic writer has been taken over by Sholokhov, the author of the epic narrative *Quiet Flows the Don* who sought to create a *War and Peace* of the twentieth century. The most recent claimant of the Tolstoy role is A. Solzhenitsyn, who is trying to jump even higher than Sholokhov, writing a multi-volume series of historic novels about the Russian Revolution in 1917, called *A Red Wheel*. Being heavily centralised, Russian spiritual culture opposes the standard norms of European-style civil society, built on private interests of the individual. Myths about the universal nature of Russian spiritual culture leave no room for subcultures because their private, partial, individual nature undermines the pretension of Russian culture to be universal. The tension between the concept of *Russkii Narod* based on universality (as an ethnic identity) and Russian society based on difference poses many problems for contemporary Russians who are now forced to make individual choices based on political preference, class position and so on. Thus, in the past, the very particularity of Russian culture and of all Russian people did not allow a cultural geography to manifest itself on Russian soil. Due to the character of its methods, aimed at the search for differences, geography as a discipline appeared to be at odds with the very nature of Russian culture. B.B. Rodoman once described geography as the "naturally-born anti-soviet discipline", because it insisted on giving worthwhile meaning to local identities thus opposing the very idea of the totalitarian state. In addition, it would not be an exaggeration to see regional cultural geography not only as an anti-Soviet but also an anti-Russian discipline, taking into account the bias of Russian culture towards universality, as mentioned above.[3]

In pre-revolutionary Russia the majority of geographical descriptions were addressed not to the Russian historical heartland but to the newly colonised and annexed areas. This underlines once again the extrovert character of Russian culture. Two major works, published at the end of the nineteenth century under the supervision of the head of the Russian Geographical Society P.P. Semenov-Tyan-Shanskiy: *Picturesque Russia* (1896–1904) and *Russia: the Complete Geographical Description of Our Fatherland: a Desktop Road Book for Russian People* (1902–06) served more as an exception rather than a rule. But even they show some typical Russian attitudes towards space. Both works were written from the point of view of the aristocracy and civil servants who wished to learn more about their country (at last). It is for this reason that these works, while being very professional, nevertheless read more like a catalogue or inventory. The participation of the authors in the culture which they were describing is hardly apparent, even if their interest in the subject cannot be denied. It is as if the disembodied description of the culture was more important for them. The greater pity is the fact that the spatial organisation of Russian society was already taken by them as given, based on the administrative division of the territory. *Picturesque Russia*, for example, provides a description of *gubernyi*, which were administrative units of the Russian Empire, but not the vernacular territorial communities of people. There was no attempt to recognise the voluntary social organisation of space.

A culture of distance

Because the image of monotonous Russian space is applied to the entire country, it excludes the possibility of a regional understanding of Russian culture; however, such an image can have geographical dimensions. As a rule the experience of the monotony of Russian culture is related to latitude, and the direction of Russian expansion into Siberia. It is not without reason that A. Bitov experienced his nightmare passing the west Siberian lowland. Siberia in this case is not so much a cultural region, it is more about a goal, setting a direction for expansion. It is either an attractive point or a place to escape to – "I want to go to Siberia and the Far East, where there is plenty of money and no power at least". For Russians, Siberia has no authentic cultural basis of its own. It is determined through opposition to the Russian west. This opposition began from

the times of the first voluntary escape and forced exile from the European part of Russia. Both of them are directly related to the remoteness of Siberia, to distance. On the one hand, Siberia was an asylum for free-thinking people in Russia: peasants escaped from the serfdom (which never existed in Siberia); old-believers from the modernised Greek-like Orthodox doctrine. On the other hand, Siberia was Russia's Australia: the Tsar made it the place of exile for the Decembrists and the Bolsheviks, replacing capital punishment by proscription, and afterwards the Bolsheviks organised there the Gulag Archipelago.[4] It was to Siberia that Soviet heavy industry escaped from Hitler's bombs and Siberian oil made the Soviet reign secure against the bankruptcy of the Soviet economic model. People escaped to Siberia not to develop it, but to wait. The remoteness of Siberia from European Russia was its main characteristic, which was used or overcome. Every time the development of Siberia was temporary, as a response to something else, that is why real development never took place at all. In Siberia a place is not a locality (community), it is just a location, an interplay of latitudes and longitudes. It was Siberia that many Russian thinkers connected with notorious Russian spaciousness, which "could not form persistence in the internal struggle" (N. Fedorov) and did not favour the "development of discipline and independence in a Russian man", and did not need "intensive energy, or intensive culture" (P. Chaadaev) (Zamyatin and Zamyatin, 1994).

Not only large distances, but the monotony of the natural landscape also contributed to the formation of such images. The vegetation is much less diverse than somewhere as small as Costa Rica. Siberia lies too far to the north. As the climate becomes more continental, the zone of fertile steppes, which goes along the fiftieth parallel, gradually declines; near Barnaul it becomes interrupted and soon disappears completely, with the exception of small steppe areas in Tyva, Buriyatia and near Abakan. Here the epithets "extreme" and "far" are often met. The aesthetics of the "world's end" are characteristic for the image of the region, where the land of living things gives way to the ice kingdom, and a person can disappear in endless space. As V. Podoroga noted, in Siberia Russia's Empire becomes "a part of nature". Almost one-third of Siberian land lies beyond the Arctic Circle; this a subarctic natural zone *par excellence*, where snowstorms and blizzards can be expected at any time of year. One can even mix the last spring snow with the first autumn snow; noon can be expected in the middle of the polar night, and the sun shines at midnight. Just a narrow inhabited strip stretching alongside the Trans-Siberian Railroad, mostly to the south of it, can be recognised as developed land. Three-quarters of Siberia's population and the majority of the main Siberian industries lie in this narrow belt. To the north of Trans-Siberian Railway the density of the population is negligible – 1–2 per square kilometre. The image of an archipelago, successfully created by A. Solzhenitsyn, corresponded not only with the distribution of labour camps in Siberia, but also with the geographical distribution of regular non-prison settlements in Siberia. In Siberia, European Russia is often described as the "Big Land" or "Continent". From a pragmatic viewpoint, the development of Siberia has been a continuous cycle of miscalculations and failures. The initial development was pure conquest, subjugation and robbery of the indigenous population. Colonisation was aimed at fur – the only "hard currency" used by Russia in trade with foreign countries. Fur was collected as a tribute from the local population. As a rule Russians took no part in hunting sable, they just collected the furs from the indigenous population. This reduced the positive influence of the Siberian campaign on Russian culture (Kantor, 1994). There were few people who rushed for free Siberian lands. The interests of most settlers were focused on the Volga area and did not extend to the steppes of western Siberia. Siberia was perceived by peasants as too distant and unsafe. It was the Russian north (with its severe climatic conditions) that supplied settlers for Siberia.[5] It is not without reason that first Russian colonisers were from the Russian north: S. Dejnev was born in Veliki Ustug, and E. Khabarov born in Vologda. According to estimates, up until the nineteenth century two-thirds of settlers in Siberia came from the northern Urals, one-quarter from Russian Pomorie (Arkhang'esk), with only one-tenth from the Volga area, and almost nobody from central and southern Russia (Pokishshevskiy, 1951).

The next, agricultural, wave of colonisation appeared to be forced by the lack of agricultural supply for the remote settlements. It was expensive to import the required bread and food to Siberia. Siberia's agricultural lands were scanty and did not promise big returns. The exception was the south-western parts of Siberia around Tobolsk and Barnaul where four-fifths of Siberia's rural population were concentrated (Pokishshevskiy, 1951). The industrial development of Siberia was unsuccessful as well. The Russian Empire expended its main geopolitical forces in south-eastwards and westerly directions. The Siberian branch of the Russian colonisation stream seemed to be a far from primary concern of Russian state and its strategic axes.

With the onset of industrialisation, Siberia began to attract more attention. However, up to the 1930s the progress of industrial development in Siberia was sluggish. The relatively well-developed southern belt of Siberia from Barnaul up to Irkutsk contributed only 1.3% of the value added of the industrial sector in the Russian Empire. Mining activities led to the creation of a few settlements which soon fell into decay. The volume of minerals extracted was negligible. Before the October Revolution (1917) coal was mined just to supply the Trans-Siberian Railroad (0.8 million tonnes). The blue-collar population in large Siberian cities was also negligible. On the eve of the twentieth century Irkutsk, with the population over 50 000, had just 400 blue-collar workers registered. The industrial development of Siberia was not aimed at the development of the region, rather it was a struggle with space and nature for the sake of mineral resource extraction. The geopolitical interests of Russia resulted in the treatment of the immense space of Siberia as an obstacle on the way to the Far East. The construction of the Trans-Siberian Railway, according to the concept of the project, was not aimed at the development of the region, its agriculture, nor its industry. First of all the project contributed to the interests of overcoming the distance from Europe to the Far East. Communication with the eastern edges was more important than the development of inland regions of Siberia. In many areas the railroad passed much more to the north than was needed for the development of Siberia. Unlike a regular railroad, the Trans-Siberian did not immediately acquire a network of small tributaries to inspire nearby regions with life. But even when later this did happen, the agricultural regions of Siberia did not have an opportunity to sell wheat on the European Russian market; the central authorities saw this as an economic threat to European Russia, instead of thinking of it as a success of development.[6]

The development of Siberia began in earnest during the Soviet period. The isolationist policies of the Soviet Union government forced it to locate much of its potential in the east, as far as possible from "unsafe western" borders. This strategic goal was used to approve the construction of the Baykal–Amur railway (BAM) which goes from Ust'-Kut to Komsomol'sk-on-Amur, providing access for central Siberia to the Pacific Ocean. Notwithstanding that an army of cheap labour recruited by Komsomol was used for the construction of the railroad, the average cost of 1 km of the railroad (3 million roubles in 1989 prices) was 10 times as much as a similar rate for regular railways.[7] The region traversed by the BAM service zone is supposed to be rich in natural resources and could theoretically make a significant contribution to the national economy. In reality, at present, the costs of developing these resources are so high as to make the extraction of most of the mineral deposits in this area unprofitable. Recently it was revealed that the geological data (describing the resource potential of the region) used to support the construction of the railroad were partially falsified.

Siberia has not become a distinct cultural region of Russia; instead it has become a national theme. The romance of conquering the wilderness and the practice of compulsory labour has meant that a significant part of Russian society has run the "Siberian gauntlet". Siberia, as well as the Second World War, has touched every Soviet family. Some Russian and Soviet writers have promoted the "nationalisation" of Siberia. The literary image of Siberia paradoxically incorporates two major themes: first, the construction of a Siberian identity through an opposition to the centre and European Russia which means a separation from three-quarters of the Russian population living in the European part of Russia; second, an attempt to present a vital image of a

Siberian as the epitome of the truly Russian person, an authentic representative of the ethnic Russian people (but not the multinational population of Russia).[8]

In what way is this paradoxical myth being created? Where does the pretension which enables a Siberian to speak on behalf of the whole Russian people come from? It seems as if national consciousness cannot tell space from society. Space is replacing society. Spatial monotony becomes a synonym for the unity and indivisibility of Russian culture. The identification and then replacement of the low share of the population living in Siberia by the lion's share of Russian territory that falls within Siberia entitles Siberia to be considered an authentic Russian land. The essence of a "culture of distance" is in such a replacement.

Russia as multinational state

The image of Russia as a multinational state has been colourfully formulated by the Russian writer Alexander Kuprin: "Yes, if you want to know, there is no Russian people at all. And no Russia at all! . . . There are few million square kilometres of the space and hundreds of nationalities quite different one from the other, there are thousands of languages and a set of religions. And nothing in common, if you want to know" (quoted in Zamyatin and Zamyatin, 1994).

Russia plays the role of a "box" for the peoples. Regionalism in such a space is not possible, but not because of the universal and indivisible character of Russian culture, as it was within the framework of the cultural monotony concept. The very nature of difference in such a statement appears to be different. It is possible to be distinguished by only ethnic characteristics. Race beats space. This representation of Russian culture is reflected, for instance, in the modern revival of Cossack culture. Cossacks, consisting basically of ethnic Russians, have applied for the status of a separate non-Russian ethnic group (people). Cossacks have already achieved their inclusion in the so-called all-Russian list of oppressed and deported peoples. They would like to be able to change the record of nationality in their passports from "Russian" to "Cossack", and to include Cossacks in the so-called Peoples' Federation of Russia (Matiunin, 1994). It is important to note that in its very nature the regional community of the Cossacks, born in the southern Russian steppes, transforms cultural attributes of local life into ethnic ones, as well as they do with their territory, trying to transform it into an autonomous region. An aspiration to ethnicise regional contrasts in Russia comes not only from Cossacks. There are proposals for the creation of a so-called Russian ethnic republic, along with all the others, in order to equalise the status of the ethnically Russian areas and non-Russian republics. The idea of federalism in Russia is frequently interpreted not as the form of territorial democracy, but as a means for solving national (ethnic) problems. A similar phenomenon has been observed by D. Sidorov in relation to the Orthodox Church in Russia. According to his research, the Ukrainian Orthodox Church does not consider itself as a regional branch (*eparkhia*). It seeks status as a special type of Orthodoxy, and is establishing its own branches not only in Ukraine, but even in ethnically Russian areas in Russia, competing with the Russian Orthodox Church, of which until recently it was a part.

On the basis of the discussion above, it is possible to conclude that *the people* (*narod*, an ethnic group) category is more universal than the regional community. The status of the people can be compared to the state. Unlike the region, the people cannot be recognised as a subculture. That is why the claim for ethnic (not just cultural or social) status in Russia becomes a more important, meaningful and preferable way of expressing regional identity. Such a claim cannot be localised, it is being transformed into a diaspora – a sort of network, according to Sidorov, acquiring a ubiquitous character covering a large area. Scale isomorphism, the reproduction of state-like institutions at various scales within Russia, appears to be an expression of such a mentality. The trappings of state identity (president, administration, legislative assembly, etc.) are being used not only in relation to Russia as a whole, but to her regions as well. The regions are beginning to perceive themselves as completely equal to a whole, creating the threat of the disintegration of the whole. On the other hand, the numerous claimants and their aspirations to be ubiquitous across Russian space reduces the probability of state collapse, removing the conflict from one of territorial claims to social–political debate.

Culture of difference

Just as the image of Russia as a culturally monotonous space has geographical coordinates, so multi-ethnicity has its own geographical dimensions. It is connected with the southern direction, the primary direction of Russian expansionism. The major events of Russian history are connected with the south. The first route, remaining in national memory (the historical axis of Russia), was that from Varangians (Scandinavia) to Greece, which crossed the European (western) part of Russia. The same axis is manifested by the drift of Russian capitals from Kiev through Moscow and Vladimir to St Petersburg. Recognised as the basis of Russian identity, the Volga river flows from the north to the south. One of the first experiences of the ethnocultural variety of Russia, described in the legendary notes of Russian merchant Afanasy Nikitin, is connected with the Volga. This direction is also manifest in the imperialist conquests of the Russian and the Soviet states during almost all of their history. The expansion organised by the Russian state was aimed at the Orient, not simply at the east. The latter lay to the south from the Russian historical core. The oriental question is central for Russia not only from a geopolitical, but also from an ideological point of view, and is closely connected with the struggle for Constantinople (Istanbul). Up until the Soviet period the desire to take possession of Constantinople (Tsargrad) underlay state ideology, enforced by support from public opinion. Wars between Russia and Turkey, if not popular with the Russian people, were at least recognised as something which had a "latent reason". From the beginning of the seventeenth century Russia conducted 10 wars with Turkey and 5 with Sweden. The Russian tsars willingly declared the possession of Constantinople as their main purpose, and Russian troops several times approached the fortress walls of Istanbul. For Russia military expansion towards Constantinople was routine, constant work, and, by historical sources, it appeared to be deprived of the overt patriotism that accompanied wars with France in 1812, Nazi Germany in 1941–45, or the Russian–Japanese War of 1905–07. Similarly, Russia tried to expand towards the north-west. Five Russian–Swedish wars and two Russian–Polish wars were the result of this expansion. As the words of a popular Soviet song "How wide is my native country" reveal, Soviet culture also recognised the southern direction as the basis for Soviet identity:

from Moscow up to most remote areas
from southern mountains down to northern seas
man passes as the owner of
his immense motherland.

In the opinion of G. Gatchev, the contact of Russians with other peoples on the southern edge of Russia appears to be a determining factor for Russian national self-consciousness:

There, on the southern edge of Russia, where the ground is getting harder and drier faster, decisions can be formulated, words can be kept, conflicts can occur. Not for nothing Rus' was getting stronger. She found her face, personality, self-consciousness at first among collisions with the dry and flame-volatile bodies of southerners (Scythians, Greeks, Polovtsians, Petchenegians, Tartars, Mongols, Turks), and not without reason, looking toward the south, bogatyr frontier posts and Russian bogatyr culture have developed; and then began to interfere with cold, iron, crystal Swedes and Germans. The collision with France in 1812 year in this sense was not predetermined and could be seen as one of history's fantasies and games (quoted in Zamyatin and Zamyatin, 1994).

Thus the north–south axis in Russian culture directly corresponded with imperialist policy in its classical understanding. Along this axis Russia became a multinational empire and accumulated an experience of dialogue with other cultures, languages and peoples. However, the fate of internal differences along this axis was repressed by colonial goals and imperialist statehood.

Regional distinctions, whatever they may be, natural landscape, economy or social life, displayed themselves along the meridian more brightly (evidently) than alongside the parallels. But they were overcome by the unifying work of Russian national culture and the Russian state. It is worth while giving two examples. But before that let us first examine the ideology of state colonisation clearly described by S. Lurie (1994):

The Idealistic image of the Russian empire involves, as I would say, a logic of excuse. The Russian empire was supposed to replace the Byzantine Empire and be the Third and the Last Rome, the unique Orthodox domain on Earth. So the Russian state was supposed to carry the Orthodox idea, acquiring new territories, extending Russian power over new lands and peoples. And that was an excuse for expansion. The power of the state was directly related, in people's minds, to the strength of the Orthodox idea. And that was an excuse for the Orthodox idea to have a strong linkage with the state. Therefore, the state idea of expansion went hand in hand with the Orthodox idea.

Thus, Russian identity started to lose its ethnic character and became more and more state-oriented in its nature. Everything which served the Russian Orthodox state was supposed to be Russian. Non-Russians, if they were Orthodox, were seen as Russian. So, according to the idealistic image of the Russian Empire: to be Russian is not about ethnicity, it is about statute. According to S. Lurie, Russians do not usually feel themselves to be masters over indigenous people. Russian people often feel comfortable with indigenous populations, recognising the right of native people to be equal to them, and in many cases Russians could easily serve, or work for, a rich native person anywhere within the Russian Empire, for example, in Central Asia or in northern Caucasus.

This is true unless it happens in non-Orthodox space. The expression of ethnic identity, opposing the Orthodox state, or just the state (the state was supposed to be always Orthodox) made Russians feel very uncomfortable because this kind of ethnic identity opposed the Orthodox one and distributed the state's homogeneity. As far as possible Russians tried to neutralise such opposition. Usually, local administrations settled by Russians were ambivalent to the ethnic characteristics of the people they controlled. All people were divided according to their loyalty to the particular administration, as "peaceful" and "non-peaceful". For example, all Caucasian peoples (of whom there are an incredible variety) were given just one name Cherkes, which was actually a name of one of the more numerous peoples in the Caucasus, or even worse, they were named Tatars, which was the synonym for barbarian in Russian. The absence of curiosity concerning ethnographic matters is reminiscent of the Puritans' approach to the landscape in New England in the seventeenth century. After their diaries were analysed by specialists, it was discovered that the Puritans left almost no word in their diaries about the landscape they were developing. The New England landscape unfolded to them as "wilderness", despite the variety of the peoples and landforms.

In sum, Russians always have been ignorant, and have held ambivalent attitudes towards ethnic problems. Russians are ethnically colourblind people, because they have never been a minority. There was no ideological programme of Russian colonisation, comparable with that of the British, accentuating the supply of enlightenment and civilisation to colonised people as their main objectives. There was no need for this, the process of colonisation was obvious, it went without saying. Numerous colonisation approaches were applied in the Caucasus, but none of them dominated the others, it was like a patchwork. This homogeneity could be achieved in hundreds of ways. There was no special doctrine of colonisation. Each case was unique.

The Oriental question

The Oriental question was geopolitically and ideologically crucial for Russia. However, notwithstanding the eagerness of the Second Rome, it seemed almost as if there was some sort of "taboo" acting against the successful capture of Constantinople. That is why the first real plan to capture the city was not developed until 1914 (during the First World War), but the mission was not accomplished. A long time before the First World War Russian tsars were declaring that the seizure of Constantinople was their main objective, but this goal was only really taken seriously by Catherine the Second. In 1833 the Russian fleet was in the Bosporus, and in the 1870s Russians were close to the walls of Istanbul, but made no attempt to attack the city. They

retreated nervously, as if they had doubts over their real intentions, and were fearful of losing their presence of mind. Why was this? Ideologically to occupy Constantinople meant to return the former Christian centre to the Christian domain. But at the same time Moscow as a seat of the Third Rome might have lost its position and role (see Chapter 3). This might not have been clearly realised at the time, but was likely a subconscious feeling. Plus, Constantinople had for a long time been the Christian capital for many eastern Christian peoples, not just Russians. Russia could not pretend to be the only possessor of Constantinople. That is why in acquiring new Christian countries, who sought to share the Byzantine legacy with Russia, Russia herself became involved more and more in ambiguous and contradictory ideologies, which resulted in the so-called "Oriental inferiority complex". Two principles clashed within the Russian imperial consciousness: from one side, religious Orthodox understanding that Christian people in the Russian Empire wherever they have come from were equal to Russians. And from the other, the understanding that the state was supposed to be homogeneous, which required systematic and permanent coercion. While force against Muslims could be approved to some extent by Orthodox Christians, violence against Christian peoples simply undermined the ideal structure of the empire and turned it into naked statism (etatism). But being deprived of the sacred function, the empire would have inevitably collapsed.

Even when the question was about Christian peoples, such as Armenians and Georgians in the Transcaucasus region, it was extremely hard for Russians to accept their similar religious nature. Here, Russians did not manage to impose their administration as successfully as they did in Muslim provinces. Both Armenians and Georgians were Christian peoples and had the same claims upon the Byzantine Empire's cultural heritage as Russians. Moreover, Armenians and Georgians took part in the military campaigns of the Russian army, for instance in the Russian–Turkic wars. Despite a positive attitude towards the Russian people and even towards being part of an empire, the local population did not always accept the Russian administration. As S. Lurie (1994) noted: "Russians felt themselves the most helpless in areas settled by civilised peoples, where Russian rule was met with satisfaction (as in Finland) or even with gladness (as in Transcaucasus region) the assimilation processes did not advance even a step forward, but still there was not any specific reason for the Russian administration to punish the local people. The region was partially conquered, but still did not become Russian." These feelings of discomfort concerning "non-Russians" forced Russian settlers to move out of the Christian Transcaucasus. They could not accept such a cultural distinction within the framework of a uniform state. This feeling of discomfort, having arisen from the inability to divide individual (cultural) and common (religious and state) identities, appeared to be dominant. These examples suggest that the subjugation of cultural differences along the southern axis was determined by Russian national character, and not only by the imperialist efforts of state.

Land-use export

Differences among cultural landscapes within European Russia, from the Black Sea to the White Sea, were subjugated in a highly peculiar way. It is a common understanding that the sequence of the natural zones from the north to the south contributed a lot to the origins of cultural differences within Russian society itself. As was clearly shown by D. Shaw (1993), zonal (spatial) non-ethnic differentiation of society could have been formed: for example conservative agrarian south, and urbanised liberal Moscow. It was hardly possible to get rid of those differences. They were manifested at the routine economic and even state level, but did not seriously touch the Russian mentality. More than that, through some latent social networks, European Russia underwent substantial unification. The Russian south started to dominate the agrarian economy across the whole of Russia, not just as the biggest producer of food, but as a successful "producer" of agricultural standards. The agrarian landscape, which is perhaps the most essential "producer" of cultural differences within the rural–urban continuum, was unified through the implementation of the "southern" land-use stereotype in

central and even northern Russia. Some standards, including technological ones (the optimal size of fields, farms and even caterpillar models), were developed in the south (Rostov, Krasnodar, Stavropol' regions) to be spread to other regions. A "southern steppe" system based on large fields was transferred to the forests of the non-chernozem where small parcels of arable land, separated one from the other by forest strips, presented the only technologically acceptable and environmentally safe agricultural system. Many areas suffered from soil erosion after the "southern system" had been applied to them. As B. Rodoman noticed, due to the export of southern land-use standards, the cultural landscape of the Russian steppes reached the southern outskirts of Moscow.

From another side, the contemporary idea of the small farm, as a part of the privatisation programme of the Russian government, has been developed in the Russian north, and is now being imposed upon the southern steppes. This is often in contradiction with the technological needs for large-scale irrigation or wheat-growing, as well as with optimal size of agricultural enterprise in the black soil belt. So the "northern" privatisation system is causing some management problems in the Russian south. Thus the imposition of external land-use (northern and southern) systems has played a unifying role for Russian culture along the meridian. However, the southern standard has won more often than not.

The Russian southern frontier dominates not only as the standard in agriculture; the more important southern norm is being extended to other spheres of the Russian way of life. It was in this region that Soviet culture, her myths, prophets and bearers were formed. It is no accident, therefore, that in the Soviet period numerous representatives of the Soviet establishment came from southern Russia. The southern region produced two winners of Nobel prizes: writer, author of the famous *Quiet Flows the Don*, Sholokhov (1962 prize), and ex-president of the former USSR M. Gorbachev (1991 prize). After the Second World War this region was the birthplace for leaders of the country as a whole. Many Soviet leaders of the Brezhnev era came from the southern edge of Russia or served a part of their career in this area. For example, N. Khrushchev was improving social engineering through the organisation of mass agricultural campaigns (corn sowing) on the virgin lands of northern Kazakhstan. In the same area in the middle of his career; L. Brezhnev supervised the virgin land development and Yu. Andropov, with whom the Brezhnev era finished, was born in the Stavropol region as was his successor M. Gorbachev. Vice prime-minister of the Russian government (responsible for ethnic and regional policy) S. Shakhrai is a Tersk Cossack by origin, and the speaker of the Council of the Federation of Russia V. Shumeiko spent a large part of his life in Rostov.

As a rule, two factors have served as a basis for this key role for the southern part of Russia: the political conservatism of the population – a rich soil for growing Communist Party élites – and an expedient geographical situation of areas such as Stavropol *kray* which encloses within its borders the resorts where the leaders of state often spent vacations. An important role was also played by the heroics of the bread campaign. Due to the annual struggle for a good harvest, the attention of the general public was often focused upon the Kuban, Stavropol and virgin land areas. Every harvest campaign newspaper was full of photos and heroes of socialist labour, and thousands of pounds of threshed grain. Pyramids of grain, as sacred as the ancient pyramids in Egypt, took their place in the ritual socialist landscape. On the basis of annual crop reports some regional economies were erected on a pedestal, while others were knocked down. The crop capacity ran the country, so did the Russian south. But, perhaps the main factor which made the south the standard for Soviet culture was social–political "averageness", based on the marginal geographical location of the region. The south is a sort of mini-model of the former Soviet Union. A type of average Soviet person exists here not as a statistical generalisation, but as a quite real phenomenon. It was here (not in Moscow) where the major Soviet political slogans and so-called labour initiatives were born and from there they were distributed all over the country. It was this region which was the test site for the Soviet state, practising gigantic projects of river water transfer, mass corn sowing, so-called communist construction projects such as the Volga-Don-2 canal construction, etc. It was here that the essence of

the Soviet attitude to law and sincere dislike of formal legal rights were perfectly expressed by the words of I. Polozkov (the Head of the CPSU in 1990): ". . . we should react according to the heart of the problem instead of doing this according to the law". This understanding ("according to the heart of the problem") – which means freedom from legal rights, instead of freedom within a legal system, is peculiar to both Soviet and Russian culture and stands as a rigid wall in the way of Western-type economic and political reforms in the south, as well as in the country as a whole. On the southern edge of the country, in the steppe area, the "Soviet Union" is now living out the rest of its days, transformed into a sort of subculture. The validity of the region as an example of the Soviet average helps to explain Gorbachev's arrival on the national political arena, as well as his political shape, successes and failures. Due to his regional background, Gorbachev could understand the country as a whole, in all its extremes; he felt and knew with whom he was dealing. He did not rely on any specific social layer; he addressed his messages to "the average Soviet person", whom he was exposed to in his native Stavropol. His policy of compromise and extreme caution in combination with courageous impulses reflected the inconsistent nature of the Russian south.

As well as Siberia, the Russian south has become a national theme, it has not simply remained a sort of regional subculture. War in Chechnya has again focused the attention of all Russia on the Russian south.

Measures of spatial difference: region versus frontier

The above-mentioned images of Russian culture have something in common: they leave no room for cultural regionalism. In the first case the possibility of discrete cultural regions is blocked by the idea of indivisibility of Russian culture; in the second by bringing regionalism into ethnicity. In both cases there is no ground for regionalism, since the only way for a geographer to represent such cultural differences in space is to put many unclosed boundaries on a map. Being unclosed these boundaries cannot distinguish one community (region) from another. The map would thus represent a continuum of borders rather than a set of discrete communities. From a cultural geography perspective, the form of spatial variation of a culture is very important. The very fact of difference does not yet make any sense. The gradual transformation of one identity into another is dependent upon the spatial context, in other words, the non-regional, gradient character of cultural differences cannot be a good reason for assuming the existence of a stable regional consciousness, easily compartmentalised into a single identity. Only the discrete nature of a cultural space can be a sufficient reason for the creation of a regional consciousness and the valid base for regional identity. It is impossible to portray Russian people (*narod*) in space as a system of regional subcultures. There are no readily identifiable "regional cultures" among ethnic Russians in Russia. Notions such as "Siberia" and "the south" are part of a nationwide identity, rather than a regional consciousness. The very term "culture" should only be used in relation to Russian society as a whole where the key distinction in terms of identity is along ethnic lines: unless you are not Russian, you are *a* Russian.

Conclusions

After the collapse of the USSR Russia found herself in a quite different cultural situation. The contraposition of Russian people with the other peoples within the former USSR did not provide opportunities for the Russian nation to look inwards on itself, to recognise internal social and cultural distinctions. Colonial and imperialist missions promoted a constant outward glance. Russia's external orientation repressed internal self-organisation. That was obviously extremely valuable for totalitarian rule which was not interested in any form of differentiation within society, in particular a spatial one. The uniform Russian culture, and her follower – Soviet culture – have reached their limits as the regulators of social unification. Soviet culture is being gradually transformed into a subculture, as it becomes the "property" of a certain social layer, and even a region, but not of society as a whole. At present culture begins more and more to function as the

basis for "social differentiation". It is being transformed from a myth into a real attribute of society. It is on its way from the spiritual world into the social one. A secularisation of culture is occurring. It is social culture which provides the possibility for the development of cultural geography in Russia and for the creation of a cultural–geographical *stranovedenie* (regional geography), not just as an intellectual exercise for academics, or the curiosity of the authorities, but as part of public self-understanding. In such a case this is not simply a process of development of a discipline, but of the self-reflection of society and its participation in the development of cultural space. A growing cultural pluralism within Russia is a key component of societal development and allows for a geography of such a society to be cultural.

Notes

1. One of the fundamental paradoxes of human geography is the contradiction between a methodological assumption (or, as a minimum, expectation) of total spatial variety of society and incompatibility of such an assumption with totality of any sort.
2. The public opinion polls in Russia show that a significant share of the citizens decline to answer questions (from 25 up to 50%).
3. The central authorities in the USSR realised the anti-Soviet nature of geography very quickly. One of the first acts of political oppression in Soviet Russia was conducted against geographers, regional experts and local functionaries who represented potential local opposition to the totalitarian regime. In the Soviet time regional geography in Russia was replaced by economic geography.
4. Exile was practised in Russia from the middle of the seventeenth century until the middle of the eighteenth century. Temporary exile to Siberia replaced capital punishment in the nineteenth century and marked the start of mass exile. In the nineteenth century the number of criminals (non-political) exiled to Siberia was over half a million. Political exile in the nineteenth century has had a loud resonance, but not in connection with the number exiled, which was considerably lower than the criminal population, but due to the popularity of the people involved (for example, F. Dostoevsky). The twentieth century has beaten all records in terms of the quantity of

political prisoners. At the beginning of the twentieth century each year about 10 000–12 000 political prisoners were deported to Siberian exile instead of being kept in prison. After the Bolsheviks took power this number dramatically increased. In communist Russia, in particular in the 1930s and 1940s, millions of Russian citizens were exiled. The notorious image of an exile country continues to contribute to the bad reputation of Siberia and stops many people from moving to the Russian east. The mobilisation of the male population achieved by exile created a huge imbalance in the demographic structure and caused social and economic problems.

5. Until Kazan – which was on the way to Siberia – was captured in 1552 by Russian troops the only possible passage to the east was through the northern Urals.

6. The fact is that on the eve of the twentieth century Siberian wheat-producers began to compete with the European farmers. This was a result of the low costs of transportation of bread: the railway practised a distance-decay tariff. Aiming to keep prices for bread in European Russia at the previous level and to protect the European market of Russia from Siberia, the Russian government introduced the so-called "Chelyabinsk turning-point". At Chelyabinsk the tariff began to be estimated back from the initial rate, that is as if the bread was carried not from Siberia, but from Chelyabinsk. Thus, the advantage of the distance-decay tariff for Siberian producers disappeared. The protectionism of European Russia was an act of a recognition of Siberian economic power. At the same time it testified to the unwillingness of the Russian government to include Siberia in the Russian market. Only 16 years later, in 1913, the Chelyabinsk turning-point was cancelled.

7. About three-quarters of works were conducted on rocks and permafrost ground. The length of the railway is about 3500 km. In 10 years 2000 km of drainage pipes were installed, 2237 bridges constructed and 3500 km of railways laid. The railway passes through seven large mountain ridges. Eight tunnels have been constructed (30.9 km), the longest one being Severomyisky (15.7 km).

8. The threat of the flooding of the hamlet of Matera caused by the construction of a large hydroelectric power station was the plot for V. Rasputin's narrative *Farewell to Matera*.

The development of resources in Siberia has been shown as a national tragedy of internal colonialism which, according to the story, appears to be a sort of state colonisation of a whole Russian people. In one of Sukshin's stories a philology professor came to Siberia to study authentic Russian folk language. The well-known former Soviet political prisoner A. Solzhenitsyn came back to Russia from the USA through Siberia, as he declared, for the sake of meeting with truly Russian people. However, in terms of number and structure of the population, there is no place in Russia less representative than Siberia. This route should be recommended to those travelling across Russia who prefer to bypass most of the country.

9. It is especially strange if one takes into account that before the revolution of 1917 Russia was conceived to be Russia only within its European part. In cultural terms at that time Siberia was conceived as non-Russia.

References

Bitov, A. (1993), *A Captive of the Caucasus*. CUP: Cambridge.

Burg, V. and Cukhchi, L. (1994), *Ocherki po geografii russkoye kultury*. Rostov.

Kantor, V. (1994), "Rossiyskoye svoebrazie: genezie i problemy", *Svobodnaya mysl'*, No. 10.

Lurie, S. (1994), "Rossiyskaya imperiya kak etnokulturnyy fenomen", *Obshestvennye nauki i sovremennost'*, No. 1: 56–64.

Matiunin, S. (1994), "The Cossack revival", *Captive Minds*, Winter–Spring: 105–18.

Pokishshevskiy, V. (1951), *Osvoenie Sibiri*. Moscow.

Shaw, D.J.B. (1993), "Geographic and historical observations on the future of a federal Russia", *Post-Soviet Geography*, 34(8): 530–40.

Semenov-Tyan-Shanskiy, P. (1896–1904), *Zhuvotskaya Rosiya*. St Petersburg.

Semenov-Tyan-Shanskiy, P. (1902–1906), *Rossiya: polnoye geograficheskoye okisaniye nashego otechestva*. St Petersburg.

Smirnyagin, L. (1995), "Russkiye v prostranstve i prostranstvo v russkikh", *Zaniye-Sila*, No. 3: 73–80.

Yampol'skiy, M. (1993), "Rossiya: kultura i subkultury", *Obshestvennyye nauki i sovremennost'*, No. 6: 58–67.

Zamyatin, A. and Zamyatin, D. (1994), *Khrestomatiya Prostranstva Rossi*, MIPOS: Moscow.

5

The republics of the Russian Federation: national territorial change

Nicholas J. Lynn
University of Edinburgh, UK

Introduction: the republics of the Russian Federation

Fifteen newly independent states emerged after the collapse of the Soviet Union. Of these post-Soviet countries, the Russian Federation is the largest and most populous, as it is in Europe as a whole. It is also the most multinational: although ethnic Russians comprise more than 80% of the total population, the last census recorded over 100 different nationalities resident in the Russian Federation. More than 20 of the non-Russian ethnic groups in the Russian Federation have, or share, their own republican "homelands". This chapter considers the experience of the "titular" nationalities of the republics during the Soviet period, and now in the newly independent Russian Federation. In the Soviet Empire the republics had only very limited regional autonomy, and the titular nationalities faced assimilation and marginalisation. This has led to demands for greater autonomy and greater freedom in post-Soviet Russia, and political relationships within the Russian Federation, especially between the republics and the federal government, have become increasingly confrontational. In turn, this has led some commentators to speculate as to whether the Russian Federation will disintegrate in the same way as the Soviet Union did (see e.g. Blum, 1994; Olshansky, 1993; Thom, 1992). Certainly, during the course of the post-Soviet transformation of the Russian state, the republics do appear to have posed a distinctive challenge to the federal government. They have been able to expand their powers and have profited from a central government which, at times, has appeared paralysed and weak.

The structure of the Russian Federation

The Russian Federation is a complex hierarchy of administrative units. According to the 1992 Federation Treaty and the 1993 Constitution of the Russian Federation, Russia is comprised of 89 different administrative units (shown on Map 5.1). There are 21 republics, made up of the former autonomous republics of the Russian Soviet Federative Socialist Republic (RSFSR) and all of the former autonomous *oblasts* of the RSFSR (except the Jewish autonomous *oblast* in the Far East). The former autonomous republics and autonomous *oblasts* of the RSFSR were areas deemed to have a distinctive ethnic or cultural character, although they were not seen to have fulfilled the special criteria needed for the establishment of union republic status. There are also 11 autonomous "formations" in the Russian Federation (the Jewish autonomous *oblast* and 10 smaller nationality districts (*okrugs*) of the RSFSR) that signed a separate Federation Treaty in March 1992 to that of the republics, and that are administered as distinctive regions within the remaining 57 regions of the Russian Federation – the "regular" *oblasts* and *krays* (which include the two city "regions" of Moscow and St Petersburg). The *oblasts* and *krays* also signed a separate Federation Treaty in March 1992, thereby reinforcing the differences between the levels of the federal hierarchy of the Russian Federation.

The federal structure of post-Soviet Russia is asymmetrical in that some parts of the federal hierarchy have more rights than others. In particular, the republics have won a number of important "privileges" from the federal authorities, especially in terms of self-government and tax and export contributions. The republics have elected presidents, written constitutions, and most have bicameral legislatures. Some republics have won special tax concessions, and have come to agreements with the federal authorities on the ownership of natural resources and the distribution of profits from their exploitation. The republics also have extra language and cultural rights. Republican constitutions, for example, declare that titular and Russian languages have equal official status in the republics. However, despite the appearance of republican "autonomy", political power in the Russian Federation remains highly centralised. The republics (according to the 1993 Russian Constitution) do not have the right to secede, and the Russian military intervention in Chechnya (Ichkeria) has

Geography and Transition in the Post-Soviet Republics. Edited by M.J. Bradshaw.
© 1997 M.J. Bradshaw & Contributors. Published 1997 by John Wiley & Sons Ltd.

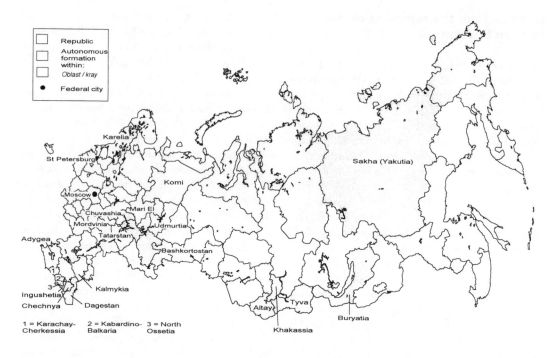

Map 5.1   *The ethnofederal structure of Russia*

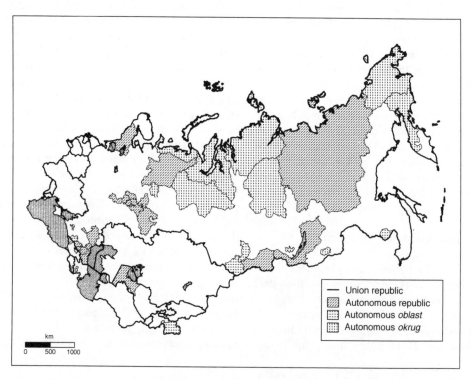

Map 5.2   *The territorial–administrative structure of the USSR*

demonstrated the lengths the federal authorities are prepared to go to preserve the country's territorial integrity. The other regions of the Russian Federation (the *oblasts* and *krays* and autonomous formations) have a very limited influence over local economic and social affairs: until the end of 1995, they did not have the right to elect their representatives directly to the upper house of the Russian parliament (the Federal Assembly).

The 21 republics only accounted for about 16% of the total population of the Russian Federation in 1989 and 29% of the territory. Their political importance does not lie in their size, but in the potential of "separatist" national movements among the titular nationality. However, Table 5.1 shows that in 1994 the titular nationality was a majority in only seven republics (in 1989 the titular nationality was the largest group in only seven republics). The demographic position in the republics is a consequence of the fact that all the titular nationalities faced decline and assimilation during the Soviet period. In 1926, the titular nationality was the largest group in at least 14 of the administrative units (Sheehy, 1991). The 1994 population figures highlight an important difference between the republics of the Russian Federation and the former union republics of the Soviet Federation. In nearly all of the Union republics there was a clear demographic basis for their demands for national independence. Republican claims for greater autonomy in post-Soviet Russia are more complex, and they more directly result from the leverage they have been able to exact from their position within the ethnofederal hierarchy. Russia has inherited a peculiar administrative-territorial legacy from the Soviet Union – a complex spatial division of power which imposes institutional limits on the process of "systemic transformation". Before it is possible to examine the new centre–republic relations that are emerging in post-Soviet Russia, it is necessary to examine the nature of the Soviet Federation in some detail.

The Soviet Federation

At the end of the Soviet period, the hierarchical national–federal structure of the USSR comprised 15 union republics (SSRs), 20 autonomous republics (ASSRs), 8 autonomous *oblasts* and 10 autonomous *okrugs* (see Map 5.2). Despite the official rhetoric, however, the fact that the Communist Party of the Soviet Union (CPSU) had a monopoly of all political power meant that the USSR functioned more like a centralised, unitary state. The rights of the union republics, and autonomous republics in turn, were subordinate to the larger "national" interests of the USSR. The decision to adopt a federal structure for the USSR was a contentious one among Lenin and the Bolsheviks, but they faced difficulties in imposing Soviet control over the borderland areas of the former Tsarist Empire during the Civil War. The root of the federal compromise in the Soviet Union lay in the Bolsheviks' desire to co-opt and undermine local interests rather than accommodate them within a "genuinely" federal structure. Soviet federalism was seen as a transitory stage that would lead to centralisation and that would facilitate the emergence of new social relations in the USSR. The distinctive character of Soviet federalism has been studied by many writers (see e.g. Smith, 1985, 1990; Kaiser, 1991; Goldman *et al.*, 1992), most of whom have emphasised the lack of historical and practical legitimacy in the Soviet federal system. At the same time, studies have also highlighted the tense *dualism* that existed in Soviet federal relations, which in practice denied any real autonomy to the non-Russian minorities, and yet provided them with the "symbolic institutions and administrative framework of autonomy" (Sakwa, 1989, p. 301). While Soviet federalism acted as a form of "socio-territorial control" (Smith, 1990) the Soviet period was also one of nation-building for selected communities within the Soviet ethnofederal system. Under Soviet nationalities policies "cultural and territorial bases of ethnicity were maintained and constitutionally safeguarded" (Smith, 1991, p. 148). Policies that nurtured the languages and cultures of different nationalities, and an internal passport system that distinguished nationality, helped to sustain and even generate nationalism in the different nationality units of the USSR. A key feature of Soviet federalism was its acceptance of *the idea of* national territoriality: the acknowledgement of some national groups'

Table 5.1 *The ethnodemographic situation in the republics of the Russian Federation*

Republic	Total population in 1989	"Titular" share of population in 1989[1]	"Titular" share of population in 1994
Karelia	790 150	10.0	10.8
Komi	1 250 847	23.3	26.3
Mari-El	749 332	43.3	40.1
Chuvashia	1 338 023	67.8	68.8
Kalmykia (Khalmg Tangch)	322 579	45.4	52.6
Tatarstan	3 641 742	48.5	48.1
Adygeya	432 046	22.1	25.2
Dagestan	1 802 188	72.4	75.1
Kabardino-Balkaria	753 531	57.6	58.1
Karachay-Cherkessia	414 970	40.9	44.7
North Ossetia	632 428	53.0	59.3
Chechnya and Ingushetia	1 270 429	70.7	99.4
Bashkortostan	3 943 113	21.9	22.9
Udmurtia	160 566	30.9	31.3
Altay	190 831	31.0	30.6
Buryatia	1 038 252	24.0	28.6
Tyva	308 557	64.3	65.6
Khakassia	566 861	11.1	9.8
Sakha (Yukutia)	1 094 065	33.4	39.6
Mordova	963 504	32.5	30.6

1. Where more than one nationality is included in the republican name, the total of both nationalities (combined) is used. "Titular" nationality in Dagestan includes: Avar, Dargin, Kumykh, Lezgin and Lak.
(Sources: Goskomstat RSFSR, 1990; Goskomstat, 1995).

territorially based claims to political recognition. Although, in reality, this principle was little more than rhetoric, the recognition of national statehood was enshrined as the fundamental basis of the Soviet Federation (Gleason, 1990).

The Soviet Federation was hierarchical and bureaucratic. Power was strictly controlled by the centre, and was concentrated in the hands of the CPSU. Federalism was used as a means of maintaining the empire inherited from Tsarist Russia, by incorporating non-Russian national groups into the Soviet state structure. Autonomous republic and *oblast* leaderships had very little influence over local affairs. The Soviet command system was administered by Soviets and ministries in Moscow, and local soviets and party units were largely responsible for administering the decisions of the central party organisation and central planning apparatus. Soviet federalism corresponded to an authoritarian model of political relationships based on the use of centralised administrative and coercive power (Mann, 1986). Republican and local governments were little more than organs of the centre and the system

was administered from the top down: Soviet federalism was a form of "federal colonialism" (Smith, 1995). Federalism in the USSR could be characterised in terms of centre–periphery relations: the centre dominated the political and economic organisation of the country.

Federalism and the collapse of the Soviet Empire

Both political and economic forces were instrumental in bringing about the collapse of the USSR in 1991. Social and economic inequality (uneven development) in the former Soviet Union between Russian and non-Russian nationalities, especially those who had been "awarded" union republic status in the Soviet ethnofederal hierarchy, was one of the most important reasons for the rise of nationalist movements in the 1980s as the freedom for political expression increased (see Ericson, 1992). As attempts at "reform communism" floundered, this nationalist challenge eventually led to the disintegration of the former Tsarist Empire, which had been claimed by the Bolsheviks and continued by the Soviet regime. The generation of nationalism in the republics of the USSR was not automatic and it was not uniform. In large part, it was a consequence of the fact that the Soviet period was one of *nation-building* for the titular nationalities of the Soviet Union's republics and autonomous *oblasts*. Despite policies intended to weaken national loyalties, the Soviet period was one of nation-building for selected communities within the Soviet ethnofederal system. Ethnicity in the Soviet Union was institutionalised both on an individual and group level, because Soviet federalism established an "inseparable link" between members of a "titular nationality" and their "territory" and their political administration (Zaslavsky, 1992, p. 71).

Since civil society was poorly articulated in the USSR, because of the monopolisation of power by the CPSU, national movements in the republics arose as a rational response to the breakdown of central state power through the institutional structure of ethnofederal units. As Gorbachev attempted to undermine the political–administrative class, which he blamed for hampering the progress of his *perestroika*, he gave more power to the institutions of the republics and autonomous republics of the Soviet ethnofederal hierarchy. Local administrative élites, who, particularly under Brezhnev, had built up their power bases in the republics, were able to exploit the shift of power from the centre without having to lose power to other groups. Breuilly (1992, p. 350) called this process "the politics of inheritance". By the time Gorbachev's proposal of a looser federal arrangement and a new Union Treaty and provoked the August 1991 coup, the momentum behind dissolving the USSR had become unstoppable.

The autonomous republics and the collapse of the USSR

The autonomous republics and *oblasts* played an important part in the process that led to the disintegration of the USSR, declaring their national "sovereignty" in 1990 as the main political actors all manoeuvred for their support. As was the case

in the union republics, the Soviet period was also one of nation-building for the titular nationalities of the autonomous republics – many of whom had not achieved national consciousness by 1917 (Goldman *et al.*, 1992). At the same time, the titular nationalities of the autonomous republics had been subjected to far greater pressures of assimilation and Russification than the titular nationalities of the union republics. Throughout the 1980s national tensions in the autonomous republics and autonomous *oblasts* of the RSFSR intensified, and increasingly political events encouraged the ambitions of republican and *oblast* leaders for greater power. The autonomous republics and *oblasts* benefited from the political battle in Moscow between the Russian President Boris Yeltsin and the Soviet President Mikhail Gorbachev, when they were courted (offered greater autonomy and self-government) by both sides.

The autonomous republics and *oblasts* made a series of "sovereignty" declarations in 1990, most of which claimed to change their status to that of union republic in the USSR in their own right, and most of which declared ownership of the natural resources located on their territory. The autonomous republics also followed the process of "presidentification" taken by the union republics. In June 1991 the Republic of Chechnya declared its total independence from the Russian Federation, a move not followed by any of the other autonomous republics and *oblasts* of the former RSFSR, which signed a Federation Treaty in March 1992 that established them as "constituent republics within the Russian Federation" (*Federativny Dogovor*, 1992). The March 1992 Federation Treaty established the regional composition of the Russian Federation that was formalised in the 1993 Russian Constitution. By the end of 1994, which witnessed an increasing centralisation of power in the Russian Federation, all of the republics had signed agreements with the federal government (Tatarstan and Bashkortostan eventually signed bilateral agreements with the central government after refusing to sign the Federation Treaty in 1992), and a new pro-Moscow regime was being installed in Chechnya (Ichkeria) by military force.

Although the Russian Federation has managed to preserve its territorial integrity, there are many problems in centre–republic relations that remain unresolved. The Federation Treaty deliberately avoided the important issues of the ownership of natural resources, the distribution of profits from exports, and decisions on taxation and subsidies. The 1993 Russian Constitution did not resolve these issues either, which were said to come under the "joint jurisdiction" of the central government and the units of the Russian Federation (*Rossiiskaya Gazeta*, 10.11.93). However, it is equally clear that there is a different dynamic at work within the Russian Federation from that which facilitated the rapid collapse of the USSR. Instead of continued national–territorial disintegration, the republics of the Russian Federation have been engaged in projects of nation-building and state formation within the Russian Federation: the republics are developing as national states *within* the Russian Federation. This has posed particular challenges for the attempt at creating a more acceptable and more "democratic" federal structure in post-Soviet Russia.

Post-Soviet nation-building and state formation

Since 1991 the Russian Federation has embarked on a determined programme of state formation. For example, Russia has established a new Federation Treaty (1992) and a new constitution (1993), it has introduced an ambitious set of economic reforms, and it has tried to consolidate the military machine it inherited from the Soviet regime. Of particular interest in this chapter, however, is the claim that the Russian President Boris Yeltsin has made since 1991, that he is trying to establish a system of "democratic federalism" in Russia. The greatest challenge to this claim has come from the 21 republics of the Russian Federation, which since 1991 have embarked on their own programmes of state formation within the Russian Federation. The republics have adopted their own constitutions and they have established their own bilateral treaties: with one another, with foreign states, and with the central government of the Russian Federation. In the economic sphere the republics have pursued increasingly independent development strategies based on their natural resources and the indus-

tries they inherited from the Soviet era. At the same time, the republics have also developed policies that are designed to encourage a "cultural rebirth" of the titular nationality. Republican attempts at state formation have been closely tied with attempts to sponsor nation-building among the titular nationalities, and it is possible to identify a process of *national state formation* in the republics of the Russian Federation This chapter considers two aspects of national state formation in the republics of the Russian Federation. First, republican attempts to cope with economic transformation, and second, their attempts at sponsoring a cultural and national rebirth and titular nationalities.

Economic development strategies

The republics inherited very different economic structures from the Soviet period, although they can be seen to fall into three main types of economy (with some overlap): natural resource-dependent economies, industrial urban economies and agricultural rural economies. First, there are the natural-resource-based economies, such as Komi, Sakha (Yakutia), Karelia, Bashkortostan and Tatarstan, that have managed to maintain a relative socio-economic advantage over other regions. Tatarstan and Bashkortostan have significant reserves of oil, and Sakha (Yakutia) produces nearly all of Russia's diamonds – as well as significant quantities of gold, coal and gas. The Komi Republic produces coal, oil, gas and timber, and Karelia has a large and influential forestry complex. The republics have made some significant attempts at restructuring the organisation of their resource industries. In Karelia, for example, republican committees have taken far greater responsibility for forestry licensing and planning, as the Soviet organisation Karellsprom has been split into separate private enterprises (see Myllynen and Saastamionen, 1994). In Sakha (Yakutia) a republican national development strategy is closely tied to a restructuring of its diamond industry. A special emphasis has been placed on increasing the level of diamond processing (cutting and polishing) that is carried out in the republic. This is also connected with the aim of training members of the titular nationality to work in the diamond-processing industry rather than continuing to import skilled Russian workers (Lynn, 1995). In particular, the resource-rich republics have made extra demands on the central government in terms of the distribution of production revenues and taxes. Komi, Karelia, Tatarstan, Bashkortostan and Sakha (Yakutia) have all claimed ownership of the resources located on their territories, and since the 1992 Federation Treaty and the 1993 Russian Constitution avoided defining natural-resource property rights, arrangements over taxes and revenues have been determined by bilateral agreement. These agreements have set quotas for the export of republican natural resources, determined the distribution of taxes between republican and central governments, and set the amount of revenues that can be retained by the republic. These special agreements put a strain on federal relations in Russia, as regions that lack strategic natural resources argue for a more "equitable" distribution of revenues and taxes.

Second, there are the more industrial republics (especially the republics of the Volga–Ural region), which have generally faced economic decline as central government reforms have led to deindustrialisation and restructuring. The extent to which republican economies have lost out during transformation is a result of both short-term supply-side problems (which have been a particular problem for industrial enterprises in the north), and longer-term structural problems that result from the collapse of the Soviet command system and the establishment of new economic relations. The industrial structures of the republics vary greatly and trying to unravel the local consequences of (and local influences on) economic reform is a very difficult task. However, it is possible to identify certain industries that are likely to fare worse during "marketisation" than others. Bradshaw and Hanson (1994), for example, suggest that the regions that were seen as "winners" under the Soviet regime (those associated with the production of steel and civilian machine-building) are likely to be "losers" during transformation. The republics with high shares of these "loser" industries include the Volga–Ural republics of Udmurtia (64.2% of the total industrial production in

1991), Mordova (46.8%), Tatarstan (35.1%), Mari-El (36.8%) and Chuvashia (33.9%) (Goskomstat, 1992).

Tatarstan and Udmurtia face particular difficulties during transformation, since they were also important centres of the defence industry in the RSFSR and have to contend with the problems of conversion (see Cooper, 1991). The leadership of the Republic of Tatarstan, however, has been robust in its dealings with the federal authorities over tax payments (and also natural-resource export quotas). From 1992 to 1994, for example, Tatarstan stopped transferring taxes to the federal authorities, although it received over 38 billion roubles in federal subventions – almost 4% of all the federal subventions paid to the regions and republics (*Finansovoye Izvestiya*, December 1993). Tatarstan was able to use locally collected federal taxes and federal subventions to fund social products, and to assist local industrial enterprises during the period of "shock therapy" in Russia. After refusing to sign the 1992 Federation Treaty, Tatarstan eventually negotiated a bilateral treaty "On the Demarcation of Areas of Responsibility and the Mutual Delegation of Powers between the Russian Federation and the Republic of Tatarstan" in February 1994 (*Rossiiskaya Gazeta*, 18.2.94). However, the exact status of federal and regional tax allocation was largely left unresolved (subject to other bilateral agreements), especially since the February 1994 Treaty formally allowed the republic to establish its own "National Bank" with local tax-raising powers.

The third type of republican economy are those that are dependent on agricultural production: Tyva, Kalmykia, the North Caucasus republics and Altay. These republics received relatively little industrial investment during the Soviet era and remain the poorest regions of the Russian economic periphery. They tend to be characterised by relatively poor living conditions and relatively high rates of natural population increase. There is also a clear ethnic division in employment patterns within these republics, with agricultural production largely the domain of the titular nationality, and the small amount of industrial production (whether it is asbestos or cobalt mining in Tyva, the petrochemical industry in several of the North Caucasus republics, non-ferrous metallurgy in Buryatia, or the forestry complex in Altay) largely the domain of Russians and other immigrants.

Two key aspects of republican economic development strategies, and two key issues in centre–republic relations, are the establishment of special fiscal regimes and republican economic protectionism. In terms of fiscal arrangements, the agricultural republics, in particular, are extremely dependent on federal subventions. In 1994, for example, federal subventions accounted for more than half of their local budgets (Lavrov, 1994). However, the agricultural republics are not alone in receiving a significant amount of subsidy from the federal treasury. In 1992 (since when data have been more contested), of the 15 units of the Russian Federation that received more money from the federal treasury in subventions than they paid to it in taxes, all but two (Irkutsk and Kamchatka *oblasts*) were republics (Smirnyagin, 1993). The republics have benefited most from the system of intergovernmental finances in Russia while it has been in flux, and are likely to continue to do so until a more transparent system is in place (see Wallach, 1994). During the shift from tax-sharing to tax allocation in Russia, the republics have been able to retain significant revenue for local budgets – although they argue that the revenue is necessary to meet their increasing social costs, as the responsibility for social services in Russia has shifted from central ministries to local government. It is also the case that the allocation of the greatest per capita subsidies to a number of republics (especially to Altay and Tyva) is an acknowledgement of the very poor socio-economic conditions that exist in these republics, which are among the worst in the Russian Federation.

Reliance on subventions, credits and federal investments provides a real limit to any "genuine" notion of economic "sovereignty" for many of the republics of the Russian Federation. In fact, it is possible to see the rhetoric of republican economic sovereignty as really being a form of "economic protectionism" – an attempt by republican leaders to minimise the effects of shock therapy. Novikov (1994) argues that local government action is greatest where republican economies lack diversification (therefore, especially in the natural-resource-based econ-

omies of Sakha-Yakutia and Komi, and the agricultural economies of Tyva and Altay), but is also prevalent in other specific republican regimes. Most notably, Tatarstan has employed a policy of "gradual transition to the market" by keeping prices low and subsidising key aspects of agricultural and industrial production (Novikov, 1994, p. 200). These republican strategies are part of the process of political–economic "inheritance" in the Russian Federation, and republican attitudes towards privatisation and foreign investment are determined by ideas about local economic sovereignty and the control of local resources. In the republics that have been most positive in asserting their autonomy (Tatarstan, Bashkortostan and Sakha-Yakutia), maintaining control over the local economy has been prioritised, and justified in terms of "national" issues and "national rights". Conversely, in the republics that have a less confrontational view of republican autonomy (Khakassia, Kalmykia and Dagestan), economic ties with the Russian Federation and central reform programmes have been more widely implemented. The "political" and the "economic" have become closely entwined in questions of republican national rights in the Russian Federation.

National and cultural "rebirth"

One of the principal demands of the republics of the Russian Federation has been "cultural autonomy". Primarily, this is a response to the reconfiguration of the Soviet Empire that the collapse of the USSR has brought about. The titular nationalities of the republics have demanded an end to the Russification and assimilation that were encouraged by both the Tsarist and Soviet regimes. The republics have established new programmes of language education and cultural expression (music, theatre and dance). But in most of the republics, demands for cultural autonomy have coincided with political demands for self-government and local economic management. This has encouraged some of the republican governments to support policies that help "regenerate" and "reconstruct" the traditions, customs and culture of the titular nation (through the construction of official historical discourses, the symbolism of statehood, "exhibition" and "festival") in order to facilitate "national autonomy".

Table 5.1 shows the fragile ethnodemographic position of many of the titular nationalities of the republics, all of whom faced decline during the Soviet period. With the end of the USSR, most of the republics have introduced special social programmes for the titular nationality. The most common have been in the sphere of national education. Even in the Komi Republic, for example, where the Komi only accounted for 23.3% of the republican population in 1989 (and only 70% of the Komi were fluent in the Komi language – according to the All-Union census), the republican leadership has introduced a specific programme for developing the Komi language. As in most other republics, the language of the titular nationality has equal status with Russian in Komi – both are official "state languages" (*Respublika Komi*, 1992). The republic has also established a republican programme for "preserving" and "developing" the Komi language, in the first phase of which Komi language classes were re-established in urban schools for the first time since the early 1960s (Fryer and Lynn, 1995). Language reforms in most republics have also meant that the local state has supported ethnic societies and publishing, and that new language faculties have been established in republican universities. Reforms may also include making the language of the titular nationality an official state language (so that all local government documents are published in both the language of the titular nationality and Russian), and demand that the republican president and key members of the republican government speak both languages. Most of the republics have also been renamed since 1991 (for example: *Kalmykia* has become *Kalmykia* (Khalmg Tangch) – *the Kalmyk state* – and Mari has become Mari-El – *land of the Mari*). For most of the republics, the introduction of language reforms has been a very symbolic process, re-establishing national rights that were "taken" from them in the 1960s and 1970s when Soviet propaganda charted the movement towards *sliyanie* (the "merger" of different nationalities into a single *homo sovieticus*) – which the titular nationalities understood to mean "assimilation".

Although language reform has been one of the most important developments in the republics, it has not been the only expression of cultural autonomy. Some of the republics have introduced an official policy of "national rebirth", that is based on "reviving" the culture and traditions of the titular nationality. This has taken a variety of forms, from constitutional changes to national holidays. In some cases this has been motivated by growing nationalism, in others by more practical motivations (in order to secure economic and political privileges). There are important differences between the republics in the extent to which nationalist issues have dominated the political and economic agenda. In Tyva, the North Caucasus republics and Tatarstan, national issues have been more important than in Kalmykia, Khakassia, Komi and Karelia, for example. Although, even in the republics where national issues have been among the most important local political questions, the lack of success of nationalist parties in local elections reveals the extent to which nationalism has not been a powerful mobilising force in the republics of the Russian Federation.

One celebration common to most of the republics is a "national day". For most this falls on either the anniversary of their 1990 declaration of sovereignty, or on the anniversary of the establishment of their first post-Soviet constitution. Republics have also adopted their own flags and state symbols, and finance state dance troupes and theatre companies. The extent to which these changes are "cosmetic" – in order to justify political and economic advantages in the Russian Federation – is a source of constant argument in the Russian Federation. Within the republics, the titular nationalities differ in their support of the policies of their local leaders, and other groups (indigenous peoples and immigrants) challenge the process of national-state formation which focuses on the rights of the titular nationality. Outside of the republics, *oblasts* and *krays*, and the federal authorities have challenged the republics' national façade, which some see as only a means of legitimising economic protectionism. In July 1993, for example, Sverdlovsk *oblast* (under the leadership of its controversial governor Eduard Rossel) unilaterally declared itself a republic. Although the decision (and others like it) were never recognised by the federal government, it sparked a brief process of republicanisation among *oblasts* and *krays* in Russia. Eduard Rossel justified the Sverdlovsk decision as a measure designed to counterbalance the predominance of the republics. It was a protest against the asymmetry of the Russian Federation, and marked the beginning of a period of greater centralisation in Russia (which included the establishment of a new and centralising Russian constitution and the decision to send troops into Chechnya-Ichkeria).

The republics and "democratic federation" in Russia

Political and constitutional changes in the republics have mirrored those at the federal level (in Moscow). Most of the republics have introduced the post of president, either directly elected, or appointed by the local parliament as "President of the Council of Ministers". Most of the republics have introduced bicameral legislatures, with a full-time lower house, and a regionally elected (often part-time) upper house. And most of the republics have introduced new post-Soviet constitutions, that make basic claims about "democracy", human rights and the ownership of property, and that declare the republic to be *a part of* the Russian Federation. Just as there was conflict between the Russian President Boris Yeltsin and the Russian Supreme Soviet in 1992–93, in several of the republics there were bitter exchanges and constitutional rows between president and parliament (in Moldova President Vasily Guslyannikov was removed and reinstated several times). And just as many of the politicians sitting in the state Duma and Federal Assembly in Moscow are former party workers and members of the *nomenklatura*, many of the republican presidents and heads of government are former chairmen (none of the republican presidents since 1991 have been women) of the Supreme Soviet of the autonomous republic or autonomous *oblast*, and the majority of local parliamentary representatives are former industrial managers or party workers.

A number of studies have highlighted the problems that have been involved in the attempt to "democratise" post-Soviet Russia (for example, see Freidgut and Hahn, 1994; Steele, 1994; Sakwa, 1995); the point to be made here is that these problems have been encountered at all levels of the federal hierarchy. Greater republican autonomy does not mean greater local democracy. And, in turn, greater local democracy does not necessarily equate with the idea of a "federal democracy" – something that the Russian President Boris Yeltsin has claimed that he has tried to establish since 1991. The greatest challenge for the Russian government since 1991 has been the establishment of a federal democracy. Although the idea of a federal democracy is clearly complex, a simple definition might be one in which "federation" is characterised as a dynamic political relationship (a spatial division of power) that allows for the articulation of regional and republican social, economic, ethnic, religious and other differences. Since the collapse of the USSR, however, federation in Russia has not facilitated these kinds of political–geographical relationships. Rather, there has been a reassertion of central power in Russia (since 1993 especially) and both the local and federal state apparatus has remained technocratic and corporatist. Federation has not facilitated democratic change in the institutions of local and federal government in Russia. Instead, federation has allowed a process whereby a reconstituted élite structure in the centre, regions and republics has been able to "inherit" power as it has been devolved following the collapse of the Soviet administrative command system. The task of creating a system of federal democracy in Russia will remain one of the greatest challenges for the centre, regions and republics of the Russian Federation.

A republican challenge?

There are clearly great differences between the republics in the kind of relationship they are trying to establish with the rest of the Russian Federation. However, even in those republics where the process of national-state formation has been particularly strong, there is very little desire for "secession": they are developing as "national states" *within* the Russian Federation. Therefore, although the process of national–territorial disintegration has continued since the collapse of the USSR, it has lost a lot of its momentum. Economically and politically, the republics would rather pursue national development strategies as (mostly) peripheral regions of the Russian Federation, than as peripheral independent states in the world system. The process of political and economic "inheritance" in the Russian Federation has largely been about controlling the organs of enhanced local power within (albeit loosened) federal state structures. Even in Chechnya (Ichkeria), where a national independence movement came to head the republican government (at a time of heightened crisis), it is unclear how popular a move towards real secession would have been (and just what "independence" would have actually meant in practice). Russia's military involvement in the republic, and the federal authorities' sensitivity to Chechen nationalist rhetoric, reveals more about the peculiarities of the relationship between the Russian state and the North Caucasus (that stretches back some 300 years) than contemporary centre–republic relations in the Russian Federation.

Viewing federal relations in Russia as some kind of antagonistic relationship between the centre and its republics is only one perception of centre–republic relations in the Russian Federation. The republics make very different demands on the federal authorities and they are trying to pursue very different relationships with other regions of the federation. There is a complex mix of different factors that are shaping federal relations in Russia. In order to analyse critically the challenge that the republics pose to the peaceful development of federal relations in Russia, it is necessary to understand the ways in which the republics are different, both from one another and from the other regions that make up the Russian Federation. Any political–territorial arrangement in Russia must certainly be sensitive to the different kinds of challenges that the republics pose.

Acknowledgements

The author would like to acknowledge the assistance provided by the Scott Polar Research Institute, University of Cambridge, in organising the research for this chapter, and also David Haddock, School of Geography, University of Birmingham, for preparation of the maps.

References

Blum, D.W. (1994), *Russia's Future: Consolidation or Disintegration?* Boulder, Colo.: Westview Press.

Bradshaw, M.J. and Hanson, P. (1994), "Regions, local power and reform in Russia", in Campbell, R.W. (ed.), *Issues in the Transformation of Centrally Planned Economies: Essays in Honour of Gregory Grossman*, Boulder, Colo.: Westview Press, pp. 133–59.

Breuilly, J. (1992), *Nationalism and the State*, 2nd edn. Manchester: Manchester University Press.

Cooper, J. (1991), *The Soviet Defence Industry: Conversion and Reform.* London: RIIA.

Ericson, R.E. (1992), "Soviet economic structure and the national question", in Motyl, A.J. (ed.), *The Post Soviet Nations: Perspectives on the Demise of the USSR*, New York: Columbia University Press, pp. 240–71.

Federativny Dogovor (Federation Treaty) (1992), Izdanie Verkhovnogo Soveta Rossiyskoy Federatsy (Publication of the Russian Federation Supreme Soviet). Moscow.

Freidgut, T. and Hahn, J. (1994), *Local Power and Post-Soviet Politics.* London: ME Sharpe.

Fryer, P.J.W. and Lynn, N.J. (1995), "National-state formation in Komi and Sakha republics", paper presented at the 1995 conference on "Geographies of Transformation" at the University of Birmingham.

Gleason, G. (1990), *Federalism and Nationalism: the Struggle for Republican Rights in the USSR.* Boulder, Colo.: Westview Press.

Goldman, P., Lapidus, G.W. and Zaslavsky, V. (1992), "Introduction: Soviet federalism – its origins, evolution, and demise", in Lapidus, G.W., Zaslavsky, V. and Goldman, P. (eds), *From Union to Commonwealth: Nationalism and Separatism in the Soviet Republics*, Cambridge: Cambridge University Press, pp. 1–21.

Goskomstat (1992), *Pokazateli ekonomicheskogo razvitiya respublik, kraev, oblastey Rossiiskoy Federatsy* (Indicators of the Economic Development of the Republics, Krays and Oblasts of the Russian Federation). Moscow.

Goskomstat (1995), *Raspredelinie naselenie Rossii po vladeniyu yaz'ykami* (Distribution of the Population of Russia by Language-Use). Moscow (Microcensus).

Goskomstat RSFSR (1990), *Natsional'ny sostav naseleniya RSFSR* (National Composition of the Population of the RSFSR). Moscow.

Kaiser, R.J. (1991), "Nationalism: the challenge to Soviet federalism", in Bradshaw, M.J. (ed.), *The Soviet Union: a New Regional Geography?*, London: Belhaven, pp. 39–65.

Lavrov, A. (1994), *Budgetny federalism v Rossii* (Budgetary Federalism in Russia). Paper by the Analytical Centre for the President of the Russian Federation.

Lynn, N.J. (1995), "A political geography of the republics of the Russian Federation", Unpublished Ph.D. thesis, University of Birmingham.

Mann, M. (1986), "The autonomous power of the state: its origins, mechanisms and results", in Hall, J.A. (ed.), *States in History*, Oxford: Blackwell, pp. 108–36.

Myllynen, A. and Saastamionen, O. (1994), "The transformation of the forest sector in the Karelian republic: a Finnish view", in Eskelinen, H., Oksa, J. and Austin, D. (eds), *Russian Karelia in Search of a New Role*, Joensuu: University of Joensuu, pp. 97–107.

Novikov, A. (1994), "K probleme ekonomicheskix osnov natsial'nogo suvereniteta v respublikax Rossiiskoy Federatsy (To the problem of the economic basis of national sovereignty in the republics of the Russian Federation)", in Drobizheva, L.M. (ed.), *Natsial'noe samosoznanie i natsionalizm v Rossiiskoy Federatsy nachala 1990 godov* (National Self-consciousness and Nationalism in the Russian Federation at the Beginning of the 1990s). Moscow: Rossiiskaya Akademiya Nauk, pp. 192–211.

Olshansky, D.V. (1993), *Alternative Scenarios of the Disintegration of the Russian Federation*. McLean, Va: The Potomac Foundation.

Respublika Komi (1992), *Konstitutsiya Respublika Komi*.

Sakwa, R. (1989), *Soviet Politics: an Introduction*. London: Routledge.

Sakwa, R. (1995), "The Russian elections of December 1993", *Europe–Asia Studies*, **47**: 195–227.

Sheehy, S. (1991), "The ethno-demographic dimension", in McAuley, A. (ed.), *Soviet Federalism: Nationalism and Economic Decentralisation*, Leicester: Leicester University Press, pp. 56–88.

Smirnyagin, L.V. (1993), "Politicheski Federalism Protif Ekonomicheskogo (Political versus economic federalism)", *Segodnya*, 25 June: 2.

Smith, A.D. (1991), *National Identity*. London: Penguin.

Smith, G.E. (1985), "Ethnic nationalism in the Soviet Union: territory, cleavage and control", *Environment and Planning C: Government and Policy*, **3**: 49–73.

Smith, G.E. (1990), "The Soviet Federation: from corporatist to crisis politics", in Chisholm, M. and Smith, D. (eds), *Shared Space, Divided Space: Essays on Conflict and Territorial Organisation*, London: Unwin Hyman, pp. 84–105.

Smith, G.E. (1995), "Federation, defederation and refederation: from the Soviet Union to Russian Statehood", in Smith, G.E. (ed.), *Federalism: the Multiethnic Challenge*, London: Longman, pp. 157–79.

Steele, J. (1994), *Eternal Russia: Yeltsin, Gorbachev and the Mirage of Democracy*. London: Faber and Faber.

Thom, F. (1992), "The cracks in the Russian Federation", *Uncaptive Minds*, **5**: 19–28.

Wallach, C.I. (ed.) (1994), *Russia and the Challenge of Fiscal Federalism*. Washington, DC: The World Bank.

Zaslavsky, V. (1992), "The evolution of separatism in Soviet society under Gorbachev", in Lapidus, G.W., Zaslavsky, V. and Goldman, P. (eds), *From Union to Commonwealth: Nationalism and Separatism in the Soviet Republics*, Cambridge: Cambridge University Press, pp. 71–97.

6
The Russian diaspora: identity, citizenship and homeland

Graham Smith
University of Cambridge, UK

This is a difficult moment for the Russian diaspora, wherever it is located in the post-Soviet states. More or less overnight, without any sense of having emigrated from their "homeland", 25 million of them, scattered throughout the 14 borderland states of what up until 1991 constituted the Soviet Empire, found themselves ethnic minorities in polities whose core nations regard themselves as entitled to cultural and political dominance. For some, especially those who moved into the borderland republics during the period of Soviet rule, such geopolitical change has left a sense of rootlessness as they seek to come to terms with the loss of an identific association with the one time Soviet homeland, with what Moscow had referred to as an integral part of the community of *Sovetskii narod* (the Soviet nation). Ill at ease with their new diasporic status, over 2 million have "returned" to their redefined cultural homeland of Russia since 1991 (*Delovoy Mir*, 2.12.95). For the overwhelming majority who have remained in the borderland states, this sudden and unplanned process of empire disintegration and new state formation has made them rethink what it means to become simultaneously a diaspora and a minority in polities regarded by their core nations as their historic homelands, and of the implications that follow from being severed geopolitically from Russia.

Any self-imagining of what it implies to suddenly constitute a diaspora will vary geographically and within each national community. What it means to become a member of the Russian-speaking diaspora in Latvia may be different from her or his counterpart in Uzbekistan. Similarly, to homogenise each community, to talk of say "Russians in Ukraine", is to risk overstating a collective label which denies at least the possibility that diasporic identities may be multiple and fragmented and not necessarily neatly coterminous with communities of mobilisation and resistance. Thus what we in effect are more likely to find, particularly at this time of social flux, is a fluidity of identities, as members of the diaspora reassess their sense of self in relation to the differing situational contexts in which they are located. For many, it may mean an identity

compatible with their new home, of accepting membership of the new polity but of retaining a sense of being culturally different. For others, the linkage might be with the external homeland, Russia, an association that may not be just culturally symbolic but which may also be geopolitical in scope, bound up with envisaging a national identity in which Russia again reconnects up with the boundaries of its pre-1917 empire. And for still others, it might be an identity with a Soviet homeland that geopolitically may no longer exist but with which there is still a desire to return to the predictability, social values and cultural supports associated with their recent past.

What, however, is going to be particularly crucial is not how the diaspora define themselves by reference to their own characteristics but in relation to others. In this regard, two *relational dimensions* are crucial to shaping their postcolonial identities. First, there is the relationship between the diaspora and the dominant core nation. As nations typically claim not only a special relationship with their historic homeland but also regard it as their exclusive property, what will be of particular significance is the place the core nation allocates to the "diaspora" within that vision of homeland. Despite acquiring statehood in 1991, these core nations still see themselves as having to secure a special place for their peoples within the cultural, economic and political life of their homelands. In order to achieve this aim, they have, to varying degrees, become "ethnocracies" or to use Brubaker's more useful term, "nationalising states", in which there is a tendency by their political élites to "see the state as an 'unrealised nation-state', as a state destined to be a nation-state, the state of and for a particular nation, but not yet in fact a nation-state" (1995, p. 114). In order to secure such a privileged place for the core nation within their homelands, new political élites have engaged in a range of policies that are designed to promote the language, culture, economic well-being and even political hegemony of the core nation. State-building practices such as this can carry far reaching consequences for a diaspora, for as Morley

Geography and Transition in the Post-Soviet Republics. Edited by M.J. Bradshaw.
© 1997 M.J. Bradshaw & Contributors. Published 1997 by John Wiley & Sons Ltd.

and Robins note elsewhere, "it (the homeland) is about conserving the fundamentals of culture and identity. And, as such, it is about sustaining cultural boundaries and boundedness. To belong in this way is to protect exclusive and therefore excluding identities against those who are seen as 'aliens' and 'foreigners'. The 'other' is always and continuously a threat to the security and integrity of those who share a common home" (1993, p. 8). The Russian diaspora, because they are represented in the mindset of many as a cultural remnant of the Soviet Empire and as a symbol of the "the imperial nation", have often most to fear and lose from such a conception of homeland.

The second relational dimension concerns that of the diaspora and their external homeland of Russia (see Table 6.1). For its part, Russia has adopted the role of ethnic patron towards its diaspora although exactly what form this patronage should take is hotly debated within Russia, in part bound up with differing notions of how Russia envisages the future spatial extent of its own sovereign homeland (Kerr, 1995). For the so-called "empire-rebuilders", it means pursuing directly interventionist/neo-imperialist policies in order to safeguard diasporic interests, whereas for the so-called "nation-statists", who accept the permanency of the new geopolitical boundaries of Russia, protecting the diaspora ranges from moderate calls for the borderland states to respect diaspora rights to demands for the use of economic and/or political sanctions in order to safeguard their co-nationals. It is a debate which also resonates among the diaspora. While most accept the new geopolitical map, others take a different view based on the premise that the geopolitical division of the ethnically Russian settled lands is wholly unnatural. This is particularly evident among the diaspora in the predominantly Russian ethnoregions of northern Kazakhstan, north-east Estonia, Ukraine's Donbass and in Moldova's Trans-Dniester region where the desire to secede or to become part of a larger Russia still commands support (Smith, 1996b).

Even for that part of the diaspora whose associational ties with Russia are confined primarily to a cultural identity, "the politics of back home" are still likely to be of interest. This is especially so where members of the diaspora feel that they are not being treated equally, that they have become "victims" or "second-class citizens". If only as insurance against a worsening situation they are therefore likely to take more than a

Table 6.1 *The Russian diaspora in the borderland states*

	Population		Native born/immigrant	
	Number (thousands)	Percentage of population	Native born (%)	Immigrant (%)
Estonia	475	30.3	65.1	34.9
Latvia	906	34.0	41.6	58.4
Lithuania	344	9.4	38.4	61.6
Ukraine	11 356	22.1	42.3	57.7
Belarus	1 342	13.2	32.5	67.5
Moldova	562	13.0	43.3	56.7
Georgia	341	6.3	41.8	58.2
Azerbaijan	392	5.6	56.4	43.6
Armenia	52	1.6	n.a.	n.a.
Kazakhstan	6 228	37.8	46.8	53.2
Kyrgyzstan	917	21.5	45.1	54.7
Uzbekistan	1 653	8.3	48.3	51.7
Tajikistan	388	7.6	43.3	56.7
Turkmenistan	334	9.5	47.1	52.9

Data refer to the 1989 Soviet census.
n.a. = data not available.

Sources: *Natsionalnyi sostav naseleniya SSSR*. Moscow, 1991; Arutyunyan *et al.* (1992, p. 52).

passing interest in the sorts of political supports that Russia may or may not provide. Thus Russia's extra-territorial policy of granting Russians, irrespective of where they live, the right to Russian citizenship furnishes the diaspora with an institutional support even though by accepting Russian citizenship in polities which (with the exception of Turkmenistan) do not permit dual citizenship, it means forgoing citizenship of the polity in which they reside. If, however, Russia's interest in supporting its diaspora wanes, then this may well weaken the bonds of loyalty. For the time being, however, the diaspora remain of central concern in Russian foreign policy towards what Moscow refers to as its "near abroad" (*blizhnee zarubezh'e*). Proposals such as those of the Russian parliament in March 1996 to formulate a nationalities policy that places at the top of its agenda the safeguarding of interests of Russians irrespective of where they live in the post-Soviet states is a clear reminder to the diaspora that Russia still sees itself as its ethnic patron (*Pravda*, 29.3.96).

The rest of this chapter focuses on the politics of the Russian diaspora in the Baltic states of Estonia and Latvia.[1] Within the context of the post-Soviet states, they are worthy of special consideration for three reasons. First, there is the particularity of their "nationalising state policies", central to which is membership of the citizen-polity. Based on a series of decrees culminating in their respective citizenship laws of 1992 and 1994, both Estonia and Latvia opted to exclude a third of their permanent residents, made up more or less exclusively of the Russian diaspora, from being granted an automatic right to membership of the citizen-polity. This differed from the other post-Soviet states who granted citizenship, irrespective of ethnicity or length of residency, to all those permanently residing within their bounded territory at the moment of statehood declaration. Thus Estonia and Latvia have in effect taken the notion of "a nationalising state" the furthest. This chapter therefore begins by arguing that not only has this resulted in the formation of a particular type of citizen-polity – what we can categorise as an ethnic democracy – but that particular conceptions of "homeland/foreignness" in the political discourse of their respective core nations have been

central to the structuring of Estonia and Latvia as ethnic democracies. Second, despite excluding over 1.3 million of the Russian diaspora from automatic citizenship, the two Baltic states have not witnessed ethnic violence or mass-based collective action of the type that has occurred in some other post-Soviet states. Here it will be suggested that while there is evidence that a sense of grievance does exist among the diaspora in both Estonia and Latvia, it has not been translated into social action primarily because of what resource mobilisation theorists refer to as the absence of both political opportunities and political resources necessary for collective action. Finally, Russia has taken its role as ethnic patron especially seriously in Estonia and Latvia, not least because of the citizenship question. Here, in particular, the plight of one borderland diaspora community – the city of Narva in north-east Estonia – has been uppermost, in part because it has become a focus for diasporic resistance to the citizenship laws. I examine why Narva has become such a site of resistance and the form that diasporic politics has taken in the city.

The formation of ethnic democracies

As citizen-polities, Estonia and Latvia resemble what Smooha and Hanf (1992) would classify as ethnic democracies. According to these writers, an ethnic democracy, like Israel (within its pre-1967 boundaries) or Malaysia, "differs from other types of democracy in according a structural superior status to a particular segment of the population and in regarding the non-dominant groups as having relatively less claim to the state and also as not being fully loyal" (1992, p. 32). Besides ensuring that the core nation is institutionally pivotal beyond its numerical proportion within the national territory, an ethnic democracy, they argue, has two other central features: individual civil rights are enjoyed universally and certain collective rights are extended to ethnic minorities. On the basis of these criteria both Estonia and Latvia would seem to qualify as ethnic democracies. First, the hegemony of the core nation has been achieved primarily through delimiting the scope of political rights and through language laws. In both polities, automatic citizenship is limited to those who were citizens during the previous period of independent statehood (1918–40) and their descendants. For the recently settled diaspora and others, naturalisation requires a period of residency from the base year 1990, in Estonia two years plus a one year waiting period (extended in 1995 for post-1991 settlers to five years), and in Latvia 10 years, and additionally in both, evidence of linguistic competence and an oath of loyalty. Therefore the overwhelming majority of the diaspora do not have a right to participate in national elections; consequently, Estonians and Latvians are electorally over-represented in proportion to their numbers (Smith, 1994b; Karklins, 1994). Language laws, which from 1989 onwards elevated the core nation language to the official state vernacular, also ensures that language becomes a key resource for the privileging of members of the core nations within the political, social and economic life of their polities. This is also likely to have a particular bearing on the social mobility of the diaspora and in transforming their social structure. From constituting a social group made up primarily of political and administrative élites, factory managers and urban-industrial workers in which knowledge of the local languages played little or no part in determining their place within occupational structure, those without a knowledge of the official state language now face the prospect of becoming part of a new urban underclass. Indeed, it is estimated that some 60% of the diaspora in Estonia and 40% in Latvia have no knowledge of the official state language (Rose and Maley, 1994).

Second, certain civil and political rights are enjoyed universally, most notably freedom of the press, access to an independent judiciary, the right to vote in local government elections, and within certain prescribed limits as outlined in their respective national constitutions, the right of assembly and association. Moreover, as a result of the introduction of so-called "aliens' passports" to non-citizens in both countries (Latvia in April 1995 and Estonia in January 1996), the diaspora as permanent residents have the right to travel freely within the country and to travel securely in their right to return residency from abroad. As non-citizens, however, they have only limited rights to own property. Finally, certain collective rights are supported irrespective of citizenship; the diaspora communities have particular rights concerning their own language schools and newspapers, access to television programmes in the Russian language, and to organising their own cultural associations and clubs.

What, therefore, becomes central to understanding the making of Estonia and Latvia as ethnic democracies is how the nationalist politics of the core nations took shape during the period leading up to state formation in 1991 and of how representations of homeland in nationalist politics informed the citizenship debates. For the nationalist movements (or Popular Fronts) that sprang up in 1988 in both Baltic states, resecuring the statehood that they had lost following their forced incorporation into the Soviet Union in 1940 was integral to their vision of nationness and homeland (*Narodnyi Kongress*, 1988; *Narodnyi Front Latvii*, 1988). One issue that proved particularly problematic for these mass-based Popular Fronts was that within each movement differing interpretations emerged as to the place the diaspora should occupy within a future sovereign homeland.

Two particular visions of homeland emerged which we can label "civic–territorial" and "ethnic–primordialist". Although usually equated as being respectively synonymous with "Western" and "Eastern European traditions" (Smith, 1995), these two visions often emerge simultaneously in regions with a large immigrant population where nationalist movements are faced with the dilemma of making their appeal either territorial or primordial in focus, as in Catalonia and the Basque country. In adhering to a civic–territorial project, nationalism makes its appeal inclusionary in scope based on a conception of homeland which is broad enough to accommodate all those who live and work in the national territory. Usually associated with a civic tradition of pluralist democracy, it is based on a conception of justice in which it is held that all those who live and work in the same national territory should have the security of enjoying universal rights. Membership of the political homeland is therefore defined in purely voluntaristic terms by including those who want to share the national identity or by simply accepting as members all those who live in the territory, irrespective of their sharing the language and culture of the core nation or of their desire to assimilate into the homeland community. Within nationalist politics in Estonia and Latvia it was a conception of homeland most closely associated with so-called "moderates" within the Popular Fronts, including its leadership, many of whom were former members of the Communist Party.

This contrasts with "the ethnic–primordialist project" whose vision of the homeland community is defined by genealogical descent. Here the appeal to homeland is to mobilising individuals on the basis of ethnicised, and in its most demagogic form, racialised boundaries. As a totalising conception of communal boundedness, it imposes a particular interpretation of nationness predicated on the defence of primordial characteristics against alien cultural intrusion. Carried to its logical extreme, it can lead to calls for expulsion from the national homeland. This position came closest to that of the "radical nationalists" within the Popular Fronts: the Estonian Nationalist Independence Party and the Latvian National Independent Movement and their respective "alternative parliaments", the so-called "Congress Movements" (or "Citizens' Committees"), established in 1989 in both republics as alternatives to the existing Supreme Soviets. The Congress Movements were similar to "The Committees of Correspondence" which prepared the way for the American Revolution of 1776 and among the Soviet ethnorepublics were only found in Estonia, Latvia, Georgia and western Ukraine.

Both factions within nationalist politics defined and interpreted their sense of homeland primarily in relation to the Russian diaspora. While moderates chose to play down the differences between their core nation and the diaspora by attempting to promote the idea of a more multiculturalist movement whose organisation and aims were to unite their ethnically bifurcated societies in the cause of securing a sovereign homeland (*Padomju Jaunatne*, 25.7.89), it was the radical nationalists that increasingly set the citizenship agenda, forcing the moderates to compromise their position in order to continue to secure the social base of support for leading the way to nation-statehood. What

therefore proved crucial was the way in which particular representations of the diaspora as "the colonising other" – as "sociocultural threat", as "illegal migrants" and as "politically disloyal" in relation to homeland – were appropriated by the radical nationalists and used as effective political resources to legitimise their exclusion from the citizen-polity.

The first concerned a representation of the diaspora as a threat to the cultural preservation of the national homeland, of its core national culture and to the place of that national culture within the social and cultural life of the nation. Here exclusionists were heirs to an established tradition within dissident politics that had struggled against Moscow's hegemonic strategy to use immigration as a means to facilitate the russification of the non-Russian homelands. What, however, made immigration more central in Latvia and Estonia compared with the other Soviet union republics was the combination of its scale, continuousness and its transformative impact upon the socio-economic life of the republics. Following incorporation in 1940, the migration of Russians, more or less exclusively an urban phenomenon and associated in particular with the republics' rapid industrialisation, continued throughout the whole Soviet period. During the 1970s and early 1980s, when migration into the other non-Russian republic homelands was either slowing down or in reverse, for Estonia and Latvia it was accounting for about two-thirds of their respective population growth. Consequently, the Russian share of the population had increased from one-tenth in 1939 in both republics to about one-third by 1989. In the same period, the Latvian share of Latvia's population had declined from 75.5 to 51.8%, while in Estonia it was even more marked, with the pre-war core nation proportion falling from 90 to 64.7%. Thus more radically minded members of Estonia's Popular Front, in calling for "the defence of national identification and a halt to the process of assimilation" (*Narodnyi Kongress*, 1988), advocated introducing measures to restrict citizenship to the core nation and to "suppress the migration that threatens to make Estonians a minority in their own lands" (*Narodnyi Kongress*, 1988). A resolution passed by the Latvian Popular Front described the republic's

Russians as "a huge mass of badly qualified and uncultured people" who threaten "to swamp the ancestral territory of the Latvian peoples", with some delegates going so far as to advocate the repatriation of the diaspora (Rundenschiold, 1992, p. 613). On the basis of such concerns, Estonia and Latvia led the way as early as 1989 in introducing immigration controls in terms of quotas and through fining Moscow-administered industrial ministries for the importation of labour (Kionka, 1991).

The second major resource centred on the way in which post-war immigration was associated with the illegality of Soviet rule over the national homeland. In contrast to other nationalist movements, the Baltic Popular Fronts were unique in being able to reclaim a previous and protracted period of pre-Soviet political sovereignty which the Citizens' Committees seized upon as the building block for state-rebuilding and nation re-clarification. As the Baltic peoples had never recognised the *de jure* incorporation of their national homeland, it followed that immigration during the Soviet period was held to be illegal and that post-war settlers did not have any right to automatic citizenship. Thus in being able to argue that independent statehood was to be restored on the basis of a strict legal code of continuity with the pre-Second World War republic, it could therefore be argued that only citizens of that republic (and their descendants) should serve as transmitters of this continuity. Thus for the pure restorationists, it was held to be "unfair that all residents of the republics participate in the fate of the region" (*Literaturnaya gazeta*, 18.7.89). At the founding Congress of the Latvian Citizens' Committee in 1989 it was noted that there was nothing wrong with "an ethnically pure attitude towards citizenship, there should be no hypocrisy, there is nothing shameful in Latvian-like Latvia". To do otherwise would "legalise the occupation and incorporation of Latvia into the USSR" (*Atmoda*, 12.7.89).

Finally, there was also the way in which exclusionists constructed "the colonising other" as politically disloyal, as not identifying with the national homeland, and therefore as an obstacle both to the realisation of home rule and to social stability within the envisaged sovereign stage. Russian settlers were categorised as social and political agents of sovietisation associated in particular with the military–industrial complex, the Communist Party and with the military–security apparatus. Whereas advocates of the "civic–territorial line" argued that it was only through the maximisation of Popular Front support through mass membership that home rule could be achieved effectively and peacefully, exclusionists argued that the path to secession could only be secured through limiting the regional voices of the erstwhile "colonisers" (Ginsburgs, 1990). It was a strategy whose legitimacy was greatly strengthened by the way in which opposition to political change closely mirrored ethnic divisions. The formation in 1989 in both republics of the pro-Moscow Interfront Movements, made up primarily of the diaspora, was particularly important in this regard. Despite, however, the stereotyping of the Russian diaspora as an oppositional force, the Interfront organisations were never mass movements and their membership was largely confined to members of the Party *apparat* and plant managers (Smith, 1994b). The crucial March 1991 referenda on secession also showed that over a third of Russians in both Estonia and Latvia supported independence. So despite the envisaged prospects of becoming "second-class citizens" in an Estonian- or Latvian-dominated polity, it is probable that for many Russian settlers life in a Western-style economy was judged as increasingly more favourable than remaining part of a Soviet Union now no longer able to sustain living standards, essential services, or guarantee future employment.

These conceptions of "otherness" – as socio-cultural threat, as illegal migrants and as politically disloyal – proved ultimately crucial to shaping a particular type of regime-building in Estonia and Latvia. Both governments had, by 1990, endorsed "the restored citizen-state" as the building block of state formation. That Estonia was able to move faster than Latvia in formally adopting its citizenship law in 1992 was primarily linked to beginning the process of state-building earlier which resulted in national elections under the restored citizenship model in September 1992. But it was also linked to the greater willingness of Estonia's political factions to reach a compromise on the citizenship question undoubtedly made easier than in Latvia by Estonians feeling more comfortable about the greater security of their ethnodemographic weighting (Ishiyama, 1993). In contrast, the logic of the case put forward by the Citizens' Committee in Latvia at its first post-independence Congress in 1991, that the citizenship question could only be finally resolved after the holding of post-independence national elections was accepted, so delaying legislation until a year after the June 1993 elections (*Diena*, 3.11.93); in the interim Russian representation in the parliament was weakened considerably by barring the Communist Party from the political process while the *de facto* restored model of citizenship in combination with a 16-year residency qualification was to ensure that the pivotal first post-independence elections would secure an ethnically Latvian legislature.

Ethnic democracies, however, are not immune to outside pressures. For Estonia and Latvia, membership of the global system of nation-states meant that citizenship as a political issue became internationalised in two senses. First, internationalising pressures were connected with the desire by both Estonia and Latvia to escape from being "garrison states" of Russia. For its part, Russia purposely and publicly began to link the prospect of troop withdrawals to Estonia and Latvia adopting a more benevolent attitude towards its diaspora. As Yeltsin had emphasised: "Russia has no intention to sign any agreement regarding the withdrawal of Russian troops from Latvia or Estonia until these countries bring their legislation into line with international standards" (*Rossiiskaya gazeta*, 18.7.92). This was reiterated in his Vancouver speech in April 1993: troop withdrawals would be delayed until those countries ended "the persecution of minorities" (*Nezavizamaya gazeta*, 13.4.93). Although officially both the Estonian and Latvian governments refused to link troop withdrawals with citizenship matters, Russia's eventual willingness to withdraw its garrisons, successfully negotiated by April 1994, was primarily due to Western pressures concerning economic aid. This was despite the lack of negotiations between Russia and the Baltic states on issues related to the status of the settler communities (*Nezavizamaya gazeta*, 31.3.93).

Second, and more influential, were Western pressures, bound up with the desire of Estonia and Latvia to reconnect up with Europe, of being able to sell themselves as democratic states "living by modern European conditions" in order to secure the benefits of both Europe's economic market-place and from membership of its geopolitical security structures. Interstate organisations such as the Conference on Security and Cooperation in Europe (CSCE) and the Council of Europe were an especially moderating influence. While being seen by some Western human rights organisations such as Helsinki Watch as being too accommodative, none the less such international organisations did play an important role in effecting modifications to citizenship legislation specifically with regard to amending Estonia's June 1993 Law on Aliens which would have required all non-citizens to obtain residence and work permits within two years if they wished to remain in the country. Responding to criticisms of the CSCE, on 8 July 1993 the law was amended to guarantee work permits to any alien who had settled in Estonia prior to 1 July 1990 and had been registered as a permanent resident. Similarly, Western pressures also played a part in influencing Latvia's 1994 Citizenship Law which saw the abandonment of proposed citizen quotas and the residency qualification being reduced from 16 to 10 years.

The diaspora and the politics of grievance

While ethnic democracies are likely to remain tense due to what Smooha and Hanf (1992) refer to as "the contradictions inherent in the system", Yiftachel (1992) is more pessimistic concerning its long-term effectiveness as a particular type of regime formation, noting in particular that as a form of conflict management it is likely to fail precisely because it attempts to institutionalise majority domination over another homeland nation as in the case of Israel over the Palestinians. Although ethnic identity is generally more intense within homeland minorities than in immigrant/settler societies, none the less it might be expected, as in the case of the Russian diaspora in Estonia and Latvia, that political exclusion may lead to one of two reactive consequences, what Hirschman (1970) refers to elsewhere as *exit* and *voice*. One reactive effect might be exit, where individual members of the diaspora simply remove themselves from the polity, returning to the original homeland. What, however, is evident is that few have chosen this option despite Russia's extra-territorial policy on citizenship. In the 1991–94 period, only 41 782 residents of Estonia and an estimated 60 000 of Latvia "returned" to Russia (*Postimees*, 30.10.95), suggesting possibly that for most the costs of remaining outweigh the benefits of resettlement and/or that there exists a stronger sense of identific attachment to their present homeland than is often claimed. We might, however, expect that those who have chosen to stay may attempt to change their sociopolitical condition. This option, voice, would manifest itself in some form of reactive-based ethnic mobilisation as the marginalised attempt to redress their exclusion. Indeed, most theories of collective action would predict that under circumstances of political exclusion ethnic relations would become so strained as to lead those marginalised to become frustrated or discontented based upon feelings of "relative deprivation" (Rucht, 1991; Jenkins and Klandermans, 1995). Two questions are therefore paramount: first, is there a sense of "grievance" among the diaspora, and second, if so, why has its translation into collective action been limited?

Public opinion surveys of the diaspora communities in Estonia and Latvia have tended to focus on the extent to which they identify with the homeland in which they reside and their perceived relationship with their respective core nations. What these country-specific studies report is that although relations remain tense they have improved since 1991 and that a sense of identity is now detectable among the diaspora towards membership of the new states (Kirch and Kirch, 1995; *Tsentr Issledovannii Russkikh Men'shinstv v Stranakh Blishneogo Zarubesh'ya*, 1995). In order to tease out more directly the reaction of the Russian settler communities to the citizenship legislation, a public opinion survey was conducted by the author in 1993 into the settler communities in four cities in the Baltic states.[2] Of the sample surveyed, the overwhelming majority of Russians in Estonia and Latvia –

well over four-fifths – felt that the requirements of citizenship were unjust. This contrasted with only one-fifth of the sample in Lithuania where citizenship (as elsewhere in the post-Soviet states) was granted automatically to all those resident there in 1990. These findings are also comparable to the only other major comparative study undertaken in the Baltic states, by Rose and Maley (1994), later in that year. This study reveals that when asked if the government treats Russians fairly, 61% of Russians in Estonia disagreed/strongly disagreed with the statement compared with 54% in Latvia and 17% in Lithuania.

Just over half of our sample in Estonia and Latvia did, however, feel that a residency requirement was fair, no doubt reflecting, given the timing of immigration, that for most length of residency was not a major impediment to membership of the citizen-polity. Language as a criterion of citizenship was, however, considered very differently. Most in Estonia and Latvia, some three-quarters of those surveyed, "completely disagreed" with the contention that knowledge of the core nation language should be a condition of citizenship, a position not shared by respondents in Lithuania where only one-third held this view. In all three countries, however, most accepted that a knowledge of the core nation language was justifiable in particular spheres of public sector employment, notably within the professions. Among those most concerned by the citizenship legislation in Estonia and Latvia, there was a high correlation with particular socio-economic groups in which the strongest sense of deprivation can be identified with the recently arrived middle-aged settler, engaged in either manual or white-collar employment with only a rudimentary or no knowledge of the core nation language (Smith, 1994a).

So although a sense of grievance is not as high as might be expected, none the less it is still evident. It would, however, seem that grievance is not a sufficient condition to produce a mass politics of collective action. Indeed, the majority of Russians responded negatively in both states when asked if they would demonstrate in the streets over who should be a citizen (Rose and Maley, 1994, p. 45). What has therefore been conspicuous in both Estonia and Latvia has been the absence of national strikes, large-scale demonstrations, petition marches, invasions of official assemblies, planned insurrections or ethnic riots. Although Russian-based social and cultural organisations do exist, their membership, duration and effectiveness, are limited (Smith, 1994b, pp. 181–205). While recent attempts have also been made in all three Baltic states by the so-called Russian Representative Assemblies to co-ordinate the diaspora at both the national and transnational scale, their success in both mobilising constituent support and in defining a common political agenda, has so far been limited (*Diena*, 28.1.96). In other words, as resource mobilisation theorists would argue, it would seem that the political resources and opportunity structures necessary for ethnic mobilisation are not sufficient to facilitate mass-based collective action. In the case of Estonia and Latvia, we can therefore identify three perspectives that may shed particular light on why collective action has been limited: a weak sense of community, the absence of political entrepreneurs and what Tarrow (1994) refers to elsewhere as the structure of political opportunities.

In explaining this low level of reactive-based ethnic mobilisation, most commentators adhere to what we can label as a lack of post-colonial identity formation. The Russian-speaking diaspora, it is noted, are facing "an identity crisis" associated with coming to terms with the loss of a national political identity which conflated being "Soviet" and "Russian" (Kolstoe, 1995; Aasland, 1996). Finding a post-colonial sense of identity, a reconstituted form of communal relatedness, is, however, compounded by a variety of cross-cutting markers which weakens the prospects for ethnic mobilisation. This includes social divisions based on legal status (citizens and non-citizens), extent of spatial rootedness (recent settlers and historic communities), language (monolingual and bilingual speakers) and occupational upheaval among urban communities associated particularly with the transition from public to private sector employment. Cross-cutting markers such as these place individuals within differing situational contexts that can weaken the prospects of ethnic mobilisation. Of course it would be mistaken to reify the absence of an identity to explain why diasporic mobilisation

has been limited in scope. Although peoples need to have some sense of common identity to engage in collective action, it does not depend on being a neatly delimited "community of fate". After all, collective protest actions often bring together individuals with different but overlapping associations: the minimum requirements for collective action, as Oberschall (1973) reminds us, are "shared sentiments of collective repression and common targets of oppression".

What is crucial is the role of political entrepreneurs who are not only willing and able to champion grievances but who can manipulate ethnic markers and fashion interpretations of the situation of a group that builds upon symbols of meaning embodied in community to secure mobilisation (Smith, 1995). Within ethnic and nationalist movements this role is usually played by the cultural intelligentsia who also have a particular stake in protecting and ensuring the reproduction of cultural difference. As a social stratum, however, the diaspora's cultural intelligentsia is demographically small; although it flourished during the inter-war years, it was virtually wiped out following incorporation into the Soviet Union. Instead, ethnic mobilisation has had to rely on a different type of political entrepreneur, a technical–economic and administrative–managerial élite who settled in the republics during Soviet rule. It was this stratum who played a crucial part in the setting up of the Russian counter-movements in 1989 and whose platform was anti-independence and pro-Communist Party. In the post-independence period, however, this stratum has shown little interest in diasporic politics; rather than struggle to retain their occupational niches within public sector economic management or administration, some have moved over to the private sector, making up what constitutes one of the fastest growing social groups within the Baltic states, a new Russian business élite. It is a social stratum whose time-budgets and resources are channelled into promoting its own economic self-interests rather than prioritising ethnic concerns.

What, however, is likely to prove more pivotal is the role that the state and its political institutions play in moderating collective action. As Tarrow (1994) notes, movements are created when political opportunities open up for social actors. In the case of Estonia and Latvia, how the state has structured citizenship and the citizenship-related policies that it is pursuing is of crucial importance for it enables the state to define and regulate channels of access which are as likely to demobilise as to mobilise the diaspora into collective action. First, the degree of formal access to the political–administrative system is likely to act as a brake as individual incentives to engage in collective action are linked to judgements about the prospect of its success. Additionally it may well be that members of the diaspora do not engage in collective action because they have the opportunity to free-ride, to leave the job of looking after their vested interests to "third parties" who have greater access and resources to influence political outcomes. Besides assigning such a role to Western international organisations, this may include relying on more moderately minded Estonian or Latvian political party members to look after their vested interests, or more recently, for non-citizens, leaving this to the newly formed Russian political parties that were set up at the 1995 national elections in both polities.

Second, by delimiting access to institutional politics on non-ethnic criteria, the state has helped to create "insiders" and "outsiders" among the diaspora. This has facilitated political factionalism within the diaspora, between so-called "integrationists", who have chosen institutional politics as the arena to champion citizen rights, and "hardliners", who by and large operate outside the system of political parties and institutional politics and who criticise the integrationists for giving legitimacy to it. Thus both the Estonian and Latvian authorities have been willing to enter into a dialogue with the pro-integrationist Russian Representative Assemblies, but only in so far as its representatives accept the states' legitimacy to define and regulate membership of the citizen-polity, which has alienated more radically minded activists. Moreover, by the state keeping open the prospects of the settler communities as *individuals* becoming members of the citizen-polity, thus offering the possibility of improving personal social status, individual economic well-being and employment prospects, the short-term costs of being a non-citizen are weighted against the long-term benefits of the individual adhering to the status quo. This may explain why many individuals chose to invest their time and resources in becoming citizens rather than engaging in collective action. The growth in attendance at language schools, for instance, indicates that many Russians are keen to exploit the avenues that exist to becoming citizens (*Diena*, 20.2.95).

Finally and most explicitly, established state procedures exist whereby certain non-citizens can be granted citizenship without having to go through the formalities of naturalisation. A disproportionate number of the educated classes have secured access to citizenship in this way, invariably weakening the prospects of intelligentsia-led radical oppositional politics. Thus of the 465 persons in Estonia granted citizenship for "special services" following the enactment of Estonia's 1992 Law on Citizenship, the overwhelming majority were from the non-Estonian educated classes, including some members of Narva City Council, who at that time were openly questioning the legitimacy of their region's inclusion in the new Estonia (Park, 1994). In the wake of its citizenship law, similar procedures have also been established in Latvia "for those who have provided outstanding services" (*Diena*, 31.1.95).

The non-citizen city: diaspora politics in Narva

There is, however, one locality, Estonia's north-east region, centred on the Russian borderland city of Narva, where local conditions have been more successful in facilitating collective action. Besides being overwhelmingly Russian in composition, it is a city made up more or less exclusively of non-citizens. Although only a small minority in Narva (less than 10%) support the north-east's incorporation into neighbouring Russia, a majority do favour local regional autonomy (Smith, 1994b). Opposition to the citizenship and language laws is also far more notable with over four-fifths considering them unfair. Additionally, of the cities surveyed, Narva displayed the lowest proportion (40%) of those who intended to apply for citizenship of a Baltic state, with one in five indicating the likelihood of applying for Russian citizenship. This sense of grievance has also been translated into more radical forms of collective action: strikes, demonstrations, political lobbying, and calls for referenda on local autonomy. More than anywhere else in the Baltic states, Narva has therefore become a site of resistance for the diaspora against exclusion.

Narva presents a unique case where the political resources of a communal identity and urban political entrepreneurship come together with the existence of a local political opportunity structure to enable a politics of collective action to flourish. First, there exists a far more developed, inclusive and particular form of local community based on the overlapping identities of ethnicity and class. It is akin to what Gurr (1993) has called an ethno-class, an ethnically distinct people, descendants from immigrants, who occupy particular economic niches, usually of low status. In Narva, ethnicity and class mesh, reflecting a city made up overwhelmingly of Russian speakers who moved in during the Soviet period and who know little or no Estonian[3] and whose overwhelmingly industrial workforce is dependent on a handful of large state-sector manufacturing plants for their livelihood. Thus Narva's population suffers in a dual sense: as Russian-speaking settlers in which at the 1995 Estonian national elections, only 7000 of their adult population of 64 000 could vote (*Pravda*, 5.11.95), and as industrial workers living in one of the most economically deprived regions of the Baltic states. Here industry has been particularly affected by economic severance from Russia and the other CIS countries. By 1996, unemployment affected more than one in three households, the highest of any city in Estonia or Latvia (*Rahva Haal*, 15.2.96).

Second, identity politics also has a formal outlet, local urban administrative institutions, which have provided the diaspora with an opportunity structure to mobilise the locality into collective action and to challenge the centre. Indeed, such an opportunity structure has been exploited far more by the diaspora of Narva than any other Baltic city. Urban administration has therefore become an important source of institutional capacity because it provides a means by which non-citizens have access to articulating their

grievances, providing also a site around which group mobilisation can be focused. The pre-existing local Communist Party, municipal government (town soviet) and local trade union centre all provided the city's diaspora with an interconnected base for ethno-class mobilisation. During the 1988–91 period, these institutions provided an organisational springboard for the formation and activities of InterFront (Interdvizhenie). Opposed to Estonian statehood, its pro-Soviet Party leaders, drawn from Communist Party administrators, economic managers and engineers, utilised the confrontational language of both ethnic politics and class socialism, focusing simultaneously on local fears about "Estonianisation" and "marketisation" (*Sovetskaya estoniya*, 10.3.89; Hanson, 1993).

However, a detectable sea change has occurred in the agenda of municipally led local politics. Initially, local politics remained largely based on an unreconstructed Soviet-style class and ethnic politics in which local leaders were able to mobilise the local community through strike calls and referenda concerning the north-east's political status, issues which spoke directly to the city population as both "industrial workers" and as "Russian speakers". By 1994, however, the rhetoric and issues of diasporic politics had changed. First, support from Russia seemed less conducive to the cause of the region's secession. With the successful negotiation of Russian troop withdrawal from Estonia came a sense of abandonment; although voices can still be heard in Moscow concerning "the defence of rights of Russians in north-east Estonia", the Yeltsin administration has been fearful of raising the spectre of Russian hegemony in the Baltic due to the implications it would have for Western aid. Moreover, while the cause of Russian empire-rebuilding remains a central platform to growing neo-nationalist organisations in Russia, there has been little forthcoming to the north-east in the way of support – financial or organisational – to stimulate a politics of secession. Second, with the failure of the right-wing Moscow coup of September 1993, support for Soviet-style class politics in Narva took a major step back. In the local elections the following month, the old Communist Party leadership was ousted and replaced by a more compromising style of city council. Russians in Narva also seem to have accepted that in both Estonia and Russia, a market economy is here to stay and that given Estonia's more favourable economic performance since 1991 compared to Russia, the former offers the better prospect of being able to stem their people's falling living standards.

In moving away from Soviet-style class and secessionist politics, locally led municipal politics focus on a language of redistributive justice that demands more equitable treatment – more economic opportunities, effective political participation and better public services – with acceptance of living and working within a sovereign Estonia. This has entailed focusing on issues of differential regional economic development, of the need for central government to do something about local job creation and of stimulating local economic recovery through granting the region special economic status, of rectifying the high costs of fiscal taxes collected from Narva with what the local population perceive as limited benefits in return, of securing rights to local education and to conduct local affairs in Russian, and to the state providing more resources for learning the Estonian language in order to facilitate local citizenship.

Conclusion: the diaspora and political transition

This chapter has suggested that in adopting the features of an ethnic democracy, Estonia and Latvia have provided a basis for accommodating the insecurities of their core nations. Its longevity as a model of ethnic conflict management, however, is far from certain and it will increasingly come into competition with two very different contenders. On the one hand, the successful transition to a *liberal democracy* will depend upon the ease by which the diaspora become members of the citizen-polity. By July 1995, one-fifth of the diaspora in Estonia had become citizens, while in Latvia, by January 1994, just under a third of non-Latvians had acquired citizenship but only a further 1199 persons by January 1996 had joined the citizen-polity under its 1994 law (*Diena*, 24.1.96). There is, however, little evidence to suggest that

inclusion within the citizen-polity is being accompanied by a political party system where cross-cutting socio-economic cleavages depoliticise core nation/diasporic divisions. Even with an incremental increase in citizen voters at the 1995 General Elections in both Estonia (March) and Latvia (September), the diaspora has chosen to form their own political parties, reflecting the emergence of ethnic bipolarity within the party political system. On the other hand, core nation insecurity, linked particularly to fears of an empire-rebuilding Russia or to a sense that the core nation is not fully in control of its homeland, could lead to further calls for *a Herrenvolk type regime* that extols the virtues of repatriation and sociopolitical segregation. In Latvia, in particular, with its proportionately larger diaspora community, such calls are far from being a spent force (*Neatkariga*, 16.1.96). Whichever route the Baltic states go down, it would seem unlikely that the diaspora question will be resolved in the foreseeable future.

Notes

1. The rest of this chapter draws upon my article "The ethnic democracy thesis and the citizenship question in Estonia and Latvia" which appeared in *Nationalities Papers* (Smith, 1996a). I am grateful to *Nationalities Papers* for granting permission to reprint it in a revised form.
2. The public opinion survey of the Russian settler communities in the Baltic states was based on a random sample of 517 structured interviews conducted in February 1993 in the following cities:

	Total popu-lation ('000s)	Core nation popu-lation (%)	Russian popu-lation (%)
Riga (Latvia)	910.5	36.5	47.3
Daugavpils (south-east Latvia)	124.9	13.0	58.3
Narva (north-east Estonia)	77.5	4.0	85.9
Klaipeda (Lithuania)	202.9	63.0	28.2

The choice of the four localities was based on a number of criteria: first, cities with sizeable but differing types of Russian communities, including large immigrant populations; second, cities drawn from all three polities so as to compare responses to differing state policies towards citizenship; and finally, localities were territorial secession is an option (Daugavpils and Narva). The sample was weighted according to age and gender. For fuller details of the survey, see Smith (1994a).

3. According to a 1995 survey conducted by researchers at Tartu in Estonia, only 1% of the diaspora in Narva and the other two main cities of the north-east declared that they could speak Estonian fluently, 2% fairly fluently, 34% sufficiently, and 44% declared no knowledge of the language. See Tartu Ulikooli Turu-uurimisruhm (1996).

References

Aasland, A. (1966), "Russians outside Russia: the new Russian diaspora", in Smith, G. (ed.), *The Nationalities Question in the Post Soviet States*, London: Longman, pp. 477–97.

Arutyunyan, Yu. *et al.* (1992), *Russkie. Etnosotsiologicheskie Ocherki*. Moscow: Nauka.

Brubaker, R. (1995), "National minorities, nationalising states, and external national homelands in the New Europe", *Daedalus*, **124**(2): 107–32.

Ginsburgs, G. (1990), "The citizenship of the Baltic states", *Journal of Baltic Studies*, **21**(1): 3–26.

Gurr, T. (1993), *Minorities at Risk. A Global View of Ethnopolitical Conflicts*. Washington, DC: US Institute of Peace Studies.

Hanson, P. (1993), *Estonia's Narva Problem*. RFE/RL Research Institute, Research Paper, 22 April, 10 pp.

Hirschman, A.O. (1970), *Exit, Voice and Loyalty*. Cambridge, Mass.: Harvard University Press.

Ishiyama, J. (1993), "Founding elections and the development of transitional parties: the case of Estonia and Latvia, 1990–92", *Communist and Post-Communist Studies*, **26**(3): 277–99.

Jenkins, C. and Klandermans, B. (eds) (1995), *The Politics of Social Protest. Comparative Perspectives on States and Social Movements*. London: University College of London Press.

Karklins, R. (1994), *Ethnopolitics and Transition to Democracy. The Collapse of the USSR and Latvia*. Washington, DC: The Woodrow Wilson Center Press.

Kerr, D. (1995), "The new Eurasianism: the rise of geopolitics in Russia's foreign policy', *Europe–Asia Studies*, **47**(6): 977–88.

Kionka, R. (1991), "Are the Baltic laws discriminatory?", *RFE/RL Research Report*, **156**: 21–4.

Kirch, A. and Kirch, M. (1995), "Search for security in Estonia: new identity architecture", *Security Dialogue*, **26**(4): 439–49.

Kolstoe, P. (1995), *The New Russian Diaspora*. London: Hurst and Co.

Morley, D. and Robins, K. (1993), "No place like *Heimat*: images of homeland in European culture", in Carter, E. *et al.* (eds), *Space and Place, Theories of Identity and Location*, London: Lawrence and Wishart, pp. 3–32.

Narodnyi Front Latvii, Programma (1988), Riga.

Narodnyi Kongress: Sbornik Materialov Kongressa Narodnogo Fronta Estonii (1988), Tallinn.

Natsional'nyi sostav naseleniya SSSR (1991), Moscow: Financy i Statistika.

Oberschall, A. (1973), *Social Conflict and Social Movements*. Englewood Cliffs, NJ: Prentice-Hall.

Park, A. (1994), "Ethnicity and independence. The case of Estonia in comparative perspective", *Europe–Asia Studies*, **46**(1): 69–87.

Rose, R. and Maley, W. (1994), *Nationalities in the Baltic States. A Survey Study*, Public Policy Paper, No. 222, University of Strathclyde.

Rucht, D. (ed.) (1991), *Research on Social Movements. The State of the Art in Western Europe and the USA*. Boulder, Colo.: Westview Press.

Rundenschiold, E. (1992), "Ethnic dimension in contemporary Latvian politics: focusing forces for change", *Soviet Studies*, **44**(4).

Smith, A. (1995), *Nations and Nationalism in a Global Era*. Oxford: Polity Press.

Smith, G. (1994a), *Nationality and Citizenship in the Baltic States*. Report to the Institute of Peace Studies, Washington, DC.

Smith, G. (ed.) (1994b), *The Baltic States. The National Self-determination of Estonia, Latvia and Lithuania*. London: Macmillan.

Smith, G. (1996a), "The ethnic democracy thesis and the citizenship question in Estonia and Latvia", *Nationalities Papers*, **24**(2).

Smith, G. (ed.) (1996b), *The Nationalities Question in the Post-Soviet States*. London: Longman.

Smooha, S. and Hanf, T. (1992), "The diverse modes of ethnic conflict regulation in deeply divided societies", in Smith, A. (ed.), *Ethnicity and Nationalism*, Leiden: E.J. Brill, pp. 26–47.

Tartu Ulikooli Turu-uurimisruhm (Taru University Market Research Team) (1996), *Kirde-Eesti Linnaelanike Suhtumine Eesti Reformidesse Ja Sotsiaalpolitikasse* (The Attitudes of Town Residents of North-east Estonia towards Estonian Reforms and Social Policy), Tartu.

Tarrow, S. (1994), *Power in Movement, Social Movements, Collective Action and Politics*. Cambridge: Cambridge University Press.

Tsentr Issledovanii Russkikh Men'shinstv v Stranakh Blishneogo Zarubesh'ya. (1995), *Russkie v Estonii.* Moscow.

Yiftachel, O. (1992), "The concept of 'ethnic democracy' and its applicability to the case of Israel", *Ethnic and Racial Studies*, **13**(3): 389–413.

7

The emergence of local government in Russia

Beth Mitchneck[1]
University of Arizona, USA

Features of the Soviet system, such as democratic centralism, *nomenklatura* appointments and the centrally planned, command economy heavily weighted the relationship between the central, regional and local governments in favour of the central government. During Gorbachev's *perestroika*, the central government began to concede political and economic power to the republic and regional governments and to individual firms; this became known as political and economic decentralisation. The centrally initiated decentralisation quickly devolved into a drive for autonomy and sovereignty led by republics, regions and even cities. The ensuing conflict between the Soviet central government and regional governments and the economic disarray are viewed as major reasons for the collapse of the former Soviet Union (FSU). Areas of conflict included central investment, the payment of taxes and the general management of social and economic policy.

The political and economic collapse of such a highly centralised system is most keenly felt and, perhaps, best analysed at the local level. The day-to-day drama of transition occurs at the local level. For example, the economic collapse resulted in job loss and unpaid wages in particular places thus localising the impact (e.g. when a tank factory stops production this influences a group of people living and working in a particular city). The political decentralisation meant that government bodies, unaccustomed to acting independently, had to increase their initiative in order to provide the same level of goods and services to the local population that had been provided under the centralised system. In places where the local government did not take on these responsibilities, the population was at risk of going without basic services such as running water, heat and retail sales of food.

During the Soviet period, local government officials were part of two centralised hierarchies – the government and the Communist Party. The spatial structure of cities and the urban economy in Soviet Russia was shaped, in a large part, according to the needs of central institutions and the national economy rather than according to

the needs of the urban economy or local population (see literature on the "Soviet city" including Bater, 1980; Cattell, 1968; French and Hamilton, 1979; Hahn, 1988; Jacobs, 1983; Morton and Stuart, 1984; Ruble, 1990; Shapovalov, 1984, Taubman, 1973). Research has shown, however, that local governments could and did influence the central bodies that ultimately controlled local development (Bahry; 1987; Hough, 1969; Ruble, 1990).

After the breakup of the FSU, the process of the functional separation of local government from central government began. The process discussed here refers to the first two years after the breakup, 1992 and 1993. The end of 1993 saw the destruction of the representative elected local government. The rules under which local government, regional, and urban, constructed local policy changed as a result of this. The separation process confined after 1993; the recentralisation process noted in the case study below gained momentum in 1994.

The process included the reallocation of functions and responsibilities between levels of government. As such, locally elected government officials began to form urban development policies and priorities using parliamentary means (local soviets and *malyi* soviets) and executive power. (The parliamentary institutions were dissolved in autumn 1993 due to President Yeltsin's actions to subdue the uprising in Moscow). The *malyi* soviets were the operational arm of the larger local soviets and met about twice a month. One of their primary responsibilities was monitoring the budget and other expenditures. At times the parliamentary institutions worked in concert with the central government representatives at the local level (centrally appointed executives) and at times in conflict with them. Important changes with respect to local government rights and responsibilities occurred during the immediate post-Soviet period that allowed urban development to become a local policy concern of both the executive and legislative branches. This change had important implications for changing the spatial structure and direction of the urban economy. In addition, the

Geography and Transition in the Post-Soviet Republics. Edited by M.J. Bradshaw.

change also set in motion a process of regional variation of policy formation and implementation that had not existed at the same scale during the Soviet period.

The changes were caused, in part, because local government officials, both elected and appointed, began to use policy tools that resembled, but were not identical with, those of their counterparts in capitalist democracies (e.g. limited right to taxation and formation of independent expenditure policy). Local governments in Russia began to construct urban development policies that reflected their own priorities for urban development (e.g. housing construction or health expenditures). The geographical consequences of the ability to construct local policies include influencing interregional and intraregional variation of the provision of public goods and services and the location and growth of economic activity. It is now likely that some cities will have vastly better mass transit systems, educational systems or recreational facilities than other cities. The availability of employment and type of employment, such as manufacturing or services, will also vary by city according to local government policies rather than central directives. Local governments theoretically now have the potential to influence directly geographical processes such as population migration and employment patterns that will change the economic geography of Russia.

Some cities more actively form development priorities and use policy tools than others. It is very difficult at this time to evaluate which cities are more reform-minded than other cities. Few studies have been conducted, in part because of the limited availability of information and the vast data requirements to study the topic. Several studies at the regional level have suggested that fiscal pressures, the election of reform-minded politicians and the presence of a large non-Russian population contribute to increasing regional (and urban) government initiative (Mitchneck, 1995; Petrov *et al.*, 1993).

This chapter places the evolution of Russian local government within a historical context beginning with Tsarist Russia. Then using a case study of an urban district in St Petersburg, Petrodvorets (Map 7.1), the study examines the changing functions of local government, local

government initiatives in urban development and intergovernmental relations during the immediate post-Soviet restructuring period. A previous case study established that city government in Yaroslavl', an industrial city in central Russia (Map 7.2), manipulated the formal and informal budgetary systems to carve out independent expenditure policies and urban development priorities (Mitchneck, 1994). That case study also traced fiscal decentralisation to the urban level and recentralisation within the city itself (Mitchneck, 1994). The current study finds evidence of economic and political behaviour at the district level in St Petersburg that corresponds to the behaviour of the Yaroslavl' city government. Urban districts are one government level below cities and are subject to city policies. Like Yaroslavl' city, districts in St Petersburg, however, are subject to the political and economic actions of two levels of higher government – in this case the city of St Petersburg and the federal government. (N.B. The then St Petersburg mayor, Anatolii Sobchak, had actually been pressing for the creation of a US style municipal government system in Russia.) Evidence is provided in this chapter that behaviour similar to that of the Yaroslavl' city government emerged in another region and in an administrative unit of differing status. Fiscal decentralisation and the evolution of local government initiative occurred at several levels of the urban hierarchy in Russia and thus changed the nature and the shape of the urban environment in Russia.

Although the focus here is on changes in the urban environment, local government changes also influence rural places. For example, greater local fiscal responsibility leaves many rural settlements without funds for development due to the unprofitable nature of agricultural activities given preexisting Soviet price structures. On the positive side, however, rural local governments could play an extremely important role in the land reform process and privatisation of agriculture (see Chapter 8).

Historical background: Tsarist Russia and Soviet Russia

The Russian local government played a small role in town and regional development and planning

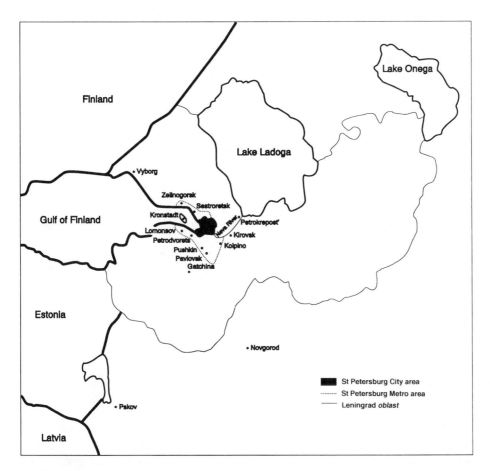

Map 7.1   *The St Petersberg region*

in Tsarist Russia and in the FSU *relative* to their Western counterparts. Local government in both Tsarist Russia and Soviet Russia traditionally had the greatest influence over the planning of economic activity pertaining to standards of living, not to industrial location. There was thus a functional separation between social welfare and economic growth. Both Tsarist and Soviet local governments were subject to a hierarchical political system in which they were clearly at the bottom of the hierarchy. The Soviet local government administered central directives from above for urban development and brought local interests to the attention of higher authorities. The Tsarist local government was also subject to central directives from above. Many similarities existed between the Tsarist and Soviet local governments in Russia including functions, ability to

initiate development activities, and central–local relations.

Pre-revolutionary Tsarist Russian local government

The Russian system of government was highly centralised and served the function of protecting central resources. Local self-government did not exist such that it could take meaningful, independent action in the area of economic development. The local government (village or town) existed essentially to support the interests of the central government and of the nobility and gentry. Historians of Tsarist Russia refer to local government bodies as local administration not as local government (e.g. Miller, 1967; Starr, 1972). This

Map 7.2   *The relative location of St Petersberg and Yaroslavl'*

may seem like a mere semantic difference; however, given local functions at this time, administration appears more appropriate than government.

During the eighteenth-century reign of Peter the Great, local government functions consisted of collecting taxes, ensuring labour service, and administering land distribution (Starr, 1972; Vinogradoff, 1979). No independent tax function existed, however, during the seventeenth, eighteenth or nineteenth centuries; fiscal activity was highly centralised (Starr, 1972). Evidence of the extreme centralisation of this government function is that from 1830 until 1870 fiscal records were not maintained at the provincial level (Starr, 1972). The lack of independent fiscal policy or responsibility at the local level indicates that local authorities did not have the means to form and implement policy different from the central government; it also demonstrates the centralised nature of the system. If legal independent action had existed, then independent accounting would have resulted.

Large-scale restructuring of government in Tsarist Russia occurred as a result of great social change, the emancipation of the serfs in 1864. The emancipation restructured the nature of local labour conditions and central–local government relations such that a new form of local government was introduced in selected provinces in 1864 – the *zemstvo*. The elimination of the previous social and political system dominated by the nobility left a gap that had to be filled in order to continue the administration of central directives. The *zemstvo* was established as a parliamentary form of government with an executive. It was defined as "a local administration which supplements the action of the rural communes, and takes cognisance of higher public wants, which individual communes cannot possibly satisfy" (Miller, 1967). The *zemstvo* was a form of rural government and government outside of the large cities.

The *zemstvo* mainly functioned in the economic realm. As before, the new form of government was responsible for tax collection and land management. Additional economic functions, however, set the *zemstvo* role firmly in the realm of regional development. The *zemstvo* was responsible for the construction of roads and canals, communication, insurance, hospitals, fire protection, primary education, enlightenment, housing for soldiers, and food sufficiency including famine warning and protection (Miller, 1967;

Vinogradoff, 1979). Vinogradoff (1979) notes that assisting the poor was the responsibility of the village communities but was shared with the Church. In addition, the *zemstvo* was supposed to provide assistance to local industry and commerce (Vinogradoff, 1979). Additional budgetary expenditures were also made on transportation, banks, the water supply, bakeries, pharmacies and saw mills (Miller, 1967). These functions are a mix of the economic infrastructure and the retail system. Some responsibility for economic growth, as well as social welfare, came to local government at this time. But the *zemstvo* had few financial resources with which to fund the expenditures. The *zemstvo* received revenues from a tax on immovable property, duties, grants, income from capital and property of the *zemstvo*, and taxes on trade and on industry (Miller, 1967). Tax arrears became a problem during this period and local indebtedness increased (Miller, 1967).

Due to the parliamentary structure of the *zemstvo* and some institutional participation in the democratic movement in the nineteenth century, a popular notion arose that the *zemstvo* represented the democratic ideal for local government. Several historians take issue with this claim and examine in detail the nature of the activism and initiative of *zemstvos* (Fallows, 1982; Manning, 1982). Manning (1982) argues that the *zemstvo* did not act independently of the central government or initiate action until the mid-1890s; these initiatives came in the form of protests to the central government over military policies of the central government (Manning, 1982). In other words, as with the original introduction of the *zemstvo* form of government, a structural change in the relations between levels of government was brought about by dramatic social or political events, the strong protests. Manning (1982) does note that during a 10-year period from 1907 until the revolution in February 1917, *zemstvos* worked together with the central government to increase the authority of local government.

Fallows (1982) examines the nature of the *zemstvo* protests to the central government Senate. This analysis provides information on both the nature of conflict between levels of government and the areas where *zemstvos* most actively sought independence. The study of protests to

the Senate shows that the majority of protests between 1890 and 1904 related to taxation (35%) and the budget (11%). In each year other than 1893 and 1897, disputes over taxes were the most numerous (Fallows, 1982). Initiatives within the formal system, then, appear to have focused on fiscal issues and perhaps on the independence of fiscal action.

Historians agree that there was a certain level of central–local conflict inherent in the structural relations of the levels of government and administration (Fallows, 1982; Manning, 1982; Starr, 1982). The above-noted study by Fallows (1982) characterises the nature of the central–local dispute as one over the distribution of tax revenues. Also, *zemstvos* worked in "harmony" with the regional level of government while maintaining adversarial relations with central ministries (Fallows, 1982). Manning (1982) describes central–local conflict during the 1905–07 period as targeted towards the dissolution of the Second Duma, the central parliament credited with radical or liberal tendencies. Central–local conflict thus seemed centred on the management of the local economy.

Soviet Russian local government

The role of local government in urban development in Soviet Russia had certain similarities of function and relations with higher levels of government with the pre-revolutionary system. The local soviet (or council) was generally responsible for the provision of goods and services to the local population and for monitoring the economic activity that occurred in its city. The monitoring function was meant to guarantee that central economic production targets were met. The Soviet system of central planning added the monitoring function as a means of ensuring that activity between ministries was co-ordinated.

The formal economic responsibilities of city and district governments were similar but not identical with those of the Tsarist local government. They were limited to the areas of public utilities, housing, mass transit, education, culture, health and the retail system (otherwise known as the communal economy and the non-productive sectors of the economy). Most urban government expenditures were made on the communal economy. In the early 1980s, these areas comprised 98% of Moscow city expenditures (Shapovalov, 1984). In other cities, sociocultural and urban economic expenditures also comprised about 98% of total expenditures (Ross, 1987).

Also like the Tsarist local government, the city and district governments were responsible for making expenditures related to urban development and the local economy, although they did not have independent decision-making authority to determine their own priorities or influence the direction of the local economy. The limitation of local authority and decision-making was significant because external decisions, made by bodies that were not responsible for local infrastructure, influenced demand for local services. Local responsibility did not include planning for economic growth. The formal sphere of influence of city government remained in those areas that influenced the development and maintenance of social overhead capital and the welfare of the population with only limited responsibility for economic overhead capital and the industrial sector (i.e. local economic growth).

Although Soviet urban governments managed the daily social and economic needs of the Soviet population, the urban governments were not independently responsible for capital investment in urban infrastructure as cities in capitalist democracies are. The urban governments were subject to decisions made by central ministries and departments about the location of economic activity. For example, in many new cities constructed by Soviet industrial ministries, decisions concerning local employment, service establishments and many other infrastructural areas for which the local governments were responsible were made without local government input (Taubman, 1973).

National priorities, rather than local priorities, were a main determining factor in the urban development of Soviet cities and districts within cities. Local governments were not entirely irrelevant to the urban development process. Research has focused on the informal networks through which urban development was managed. Findings from this research suggest that local government did more than act as the simple administrative apparatus of higher levels of government, but could and did act in its own interest or that of its community (e.g. Hough, 1969; Bahry, 1987; Ruble, 1990). Local authorities could indirectly influence urban development by lobbying for additional funds to construct needed infrastructure, such as housing or cultural facilities, or for the location of an enterprise that would increase local employment and incomes.

Local government in the post-Soviet Russia

During the transition period, local government has begun to set urban policy agendas according to its own perceived social and economic needs. A new balance between national and local priorities is being carved out at every level of government. Local governments, however, still have relatively little formal control over the formation of their budgetary revenues and the composition of budgetary expenditures, thus limiting local independence. The revenue and expenditure plans are still formed and approved by higher levels of government. Substantial evidence of growing independence in the spheres of local tax and expenditure policies and mounting initiative by local government is available, however, within the informal budgeting system and within local legislation (Berkowitz and Mitchneck, 1992). A case study of the Yaroslavl' city government shows that city government has pushed the formal boundaries of authority and responsibility into non-traditional areas, mainly due to their manipulation of the formal budgetary system (Mitchneck, 1994). Within this context, local governments can form their own priorities for urban development and the local economic activity. Yet, constant reports of local tax arrears is evidence of central–local conflict, as it was during the Tsarist period.

Despite new legislation, the formal political and economic responsibilities of local government remain firmly in the area of supporting and maintaining social and economic infrastructure as during the Soviet and even Tsarist Russian periods. The formal areas of authority were carried over into the Russian structure of local government in the post-Soviet period as evidenced by the distribution of state property during 1992.

The city and district governments retained property and enterprises involved in the communal economy and the retail system while the *oblast*, the immediate higher level of government, received property belonging to the productive or industrial sector of the economy. This gave the *oblast* access to higher tax revenue levels because theoretically the industrial sector would be income generating. (However, this does not take into account the problems associated with tax collection during periods of economic shock when production declined or ceased such that little revenue was produced.) The *oblast* also has more direct influence over economic growth.

Structural changes have occurred since 1992, the first year after the breakup of the FSU, that set the legal framework for new authority at the city level through a substantial redistribution of fiscal powers. The recent fiscal decentralisation allows for local public policy formation and the increased participation of local government in the economy. The central government passed new legislation on taxation and local government that broadened local power and authority. Russian republic laws from 1991 and 1992 give broader fiscal rights to administrative units below the republic level.[1] These laws were generally not implemented, however, until after the breakup.

Elements of independent fiscal policy are evident in an examination of the local budget in a case study of Yaroslavl' that indicate that the economic role of local government has grown and that the local government itself, not the central government, is pushing the boundaries of political and economic influence over territorial–administrative jurisdictions (Mitchneck, 1994). The increasing importance of off-budgetary revenues and expenditures,[2] the creation of one budget for the city of Yaroslavl' rather than one for each district plus the city budget, and the city-level subsidies to both the population for the social safety net and to enterprises for propping up the local economy represent significant changes in the economic role of local government during the transition period (Mitchneck, 1994).

Yaroslavl' city redefined both its economic role *vis-à-vis* central authority and its territorial power by independently forming revenues outside of the formal budget and by creating new areas of fiscal responsibility with its budgetary

and off-budgetary expenditures. The city moved from using budgetary revenues to finance limited elements of social infrastructure to financing the industrial sector of the economy with subsidies, credits and loans. The city also took on the role of financing the social safety net above and beyond the criteria set by the central government. Many similar elements are found in the Petrodvorets case study.

St Petersburg district case study

St Petersburg is located in the north-western part of Russia adjacent to Finland (Maps 7.1 and 7.2). Peter the Great built the city in the beginning of the eighteenth century to be a showcase European-style capital of Russia. It was to be "the window on the West". The city historically has been associated with progressiveness, culture and westernisation. The city is built in a swampy area along the Neva river; many small rivers and canals flow around the city thus earning it the name of the Russian Venice. Peter the Great created the city as a naval centre. During the Soviet period, the city continued to be an important site for naval and military operations and became a centre of production for the military industrial complex. During the post-Soviet period, the high concentration of military–industrial activities, including production and higher education sites, presents local officials with particular hardships during conversion to a civilian economy.

This case study analyses the development of local government in St Petersburg from 1992 to 1993 to determine the extent to which local government could independently influence urban development and participate in the local economy. These two years represent a discrete period during the immediate transition from the Soviet Union to an independent Russia. The time frame of the case study covers nearly two years of experimentation with fiscal decentralisation and the formation of local self-government immediately after the breakup of the former Soviet Union. In autumn 1993, the district soviets were abolished in St Petersburg by President Yeltsin's decree and also by the mayor of St Petersburg, Anatolii Sobchak. The district soviet deputies were locked out of their offices and the district soviet bank accounts were frozen. These actions ended the Soviet legacy of elected, representative local government at the district level in St Petersburg. The abolition of the district soviets was one chapter in the lengthy debate and struggle within St Petersburg over the establishment of "Western-style" urban government. During 1994 and 1995, several districts were consolidated to further streamline urban government.

The case study information is derived from local government documents and budgets that were collected during two field trips to Petrodvorets in June 1992 and September 1993, immediately prior to the abolition of the local soviets. The documentary data are supplemented by extensive interviews with local government officials in both the legislative and executive branches of government at the district level. Additional information was collected during 1995.

St Petersburg was divided into 24 districts in 1993, including 8 satellite cities. Petrodvorets is a satellite city (*sputnik gorod*). Satellite cities are small towns that are directly under the jurisdiction of the central city rather than the *oblast* or provincial government. They are often not contiguous with the central city and were often independent cities at some point in history. The subordination to St Petersburg is usually tied to some feature of the economy of the satellite city that gives it national priority. Petrodvorets is the home of Peter the Great's Summer Palace (also known as Peterhof) and the home of dozens of other palaces from the Tsarist period. The palaces are major tourist and historic sites, considered national treasures. There are also three military training centres there.

Petrodvorets is located along the Gulf of Finland about a 40-minute drive from the centre of St Petersburg. In order to reach Petrodvorets one drives out of the historic city centre, past the ring of the tall, modern apartment buildings still within the borders of St Petersburg, and then through Leningrad *oblast* before reaching Petrodvorets. Direct transportation between Petrodvorets and St Petersburg is limited to private cars, the commuter railroad, and the hydrofoil to the Summer Palace used by tourists. The closest metro station is about a 10- or 15-minute drive in a car. The district does have bus routes.

Petrodvorets was an independent town until 1936 when it was placed under the jurisdiction of St Petersburg. In 1989 at the first session of the newly elected district soviet, the Petrodvorets soviet passed a declaration of intent to be independent from St Petersburg and to determine the future course of development (September 1993 interview with Nikolai Ul'novich Marshin, chairman of the district soviet). The local government declared its intention to develop as a scientific, cultural, tourist and humanitarian centre. The economic base of Petrodvorets is tourism with a small industrial base consisting of five factories (several related to the military industrial complex and a large watch factory).

The weak industrial base, the importance of low-priority activities such as recreation and tourism, and the presence of the troubled military sector combine to produce less than dynamic economic conditions that are mirrored in the demographic situation of the district. The district is currently a dormitory suburb for St Petersburg with a 1 January 1993 population of 82 900, an increase of about 6000 since 1987 (77 100) but a decline since 1990 (Table 7.1). By decision of the St Petersburg city government, new housing construction over the past 20 years took place to provide housing for the overflow population from St Petersburg. Petrodvorets had been a place of net in-migration until 1991 when the district began to lose population due to out-migration (Table 7.1). The district also consistently loses population due to natural decrease – in 1992 the crude birth rate was 7 per 1000 while the crude death rate was 18 per 1000. St Petersburg as a whole also began to lose population due to net out-migration and natural decrease during this time (Mitchneck and Plane, 1994).

District government

During this first transition period, each district in St Petersburg had its own government and its own budget. In addition to the legislative branch, each district had an executive branch of government where the executive was elected by the soviet and approved by the executive or mayor of St Petersburg. These two branches of local

Table 7.1 *Demographic data for Petrodvorets*

	19.1.88	19.1.89	19.1.90	19.1.91	19.1.92	19.1.93
Population	76 800	80 200	83 500	83 800	83 800	82 900
In-migrants	8 100	5 907	4 268	3 526	3 779	n.a.
Out-migrants	5 500	4 020	3 576	4 878	4 131	n.a.
Net migrants	407	1 887	692	−1 352	− 352	n.a.
Births	989	869	821	885	618	n.a.
Deaths	1 356	1 387	1 446	1 337	1 529	n.a.
Natural change	− 367	− 518	− 625	− 452	− 911	n.a.

Source: Petrodvorets government documents.

government were, in theory, to provide checks and balance in governance. The legislative branch, being popularly elected, represented democratic self-government. The soviet was to set broad policy both independently and within frameworks mandated by higher levels of government (i.e. St Petersburg and the federal government), while the executive was to implement policies the soviet established. The *de facto* situation of governance often diverged from the theoretical; the two branches of local government did not always work in harmony and the origin of the local policies was often unclear.

Examination of changing local government functions, local economic development initiatives and intergovernmental relations yields valuable new information on the formation of local government in Russia and its role in urban development during the transition period. The formal budgets give a clear idea of the formal responsibilities and of changing functions over time of the local governments. The local initiatives in the urban economy show a self-defined economic role and are found in the use of off-budgetary revenues (surplus revenues and extra-budgetary funds), and in the formation of local policy priorities. Evidence of initiative was found in local government legislation and the resolutions from meetings of the *malyi* soviet. Information on intergovernmental relations clarifies the origins of decisions that influence the urban economy. Expenditure priorities can change according to who makes the policy. Information on the nature of relations with higher levels of government was also found in the *malyi* soviet legislation, in the form of legal protests, for example. Interviews with local officials also yielded information on intergovernmental relations.

Local government functions

Budgetary information from Petrodvorets from 1991 into 1993 shows the transition from formal Soviet budgetary practices and economic functions to Russian ones. The actual rouble expenditures are not particularly useful for analysing local government functions and priorities. Due to extremely high inflation levels during this time, rouble values are not comparable over time; thus the actual rouble amounts give less information than the percentage distribution of the expenditures. The percentage distribution indicates the relative importance of an expenditure in terms of function and priority.

In 1991, the last year of the Soviet Union, budgetary expenditures of Petrodvorets were qualitatively as well as quantitatively different than during the first two years of an independent Russia (Table 7.2). The single largest expenditure in both absolute and relative terms in 1991 was made to higher-level budgets (Table 7.2 and Figure 7.1a); 48.5% of 1991 budgetary expenditures was made to higher levels of government most likely as partial reimbursement for expenditures made by those levels on behalf of the local government. In any case, this large amount signifies the dependence that Petrodvorets had on St Petersburg. The next largest group of 1991 expenditures was made on sociocultural expenditures with the two largest being in education (21.6%) and health (12.1%). The fourth largest expenditure and last major expenditure item was on housing and the communal economy (11.2%).

The structure of the 1991 budget was relatively simple. The 1992 budget, the first in the new system, shows a restructuring of the functions of

Table 7.2 *Petrodvorets actual budgetary expenditures*

Expenditure items	Thousands of roubles			Percentage		
	1991	1992	1993, 6 months[a]	1991	1992	1993, 6 months
Housing and communal economy	8 609	84 261	266 106	11.2	23.9	19.4
Maintenance of lifeguard stations	0	0	4 296	0.0	0.0	0.3
Other communal economy	0	0	13 598	0.0	0.0	1.0
Sociocultural expenditures						
Education	16 680	130 122	671 752	21.6	36.9	48.9
Youth affairs	100	2 346	0	0.1	0.7	0.0
Culture	1 069	5 733	11 424	1.4	1.6	0.8
Health	9 335	76 368	288 699	12.1	21.6	21.0
Physical culture and sports	4	70	393	0.0	0.0	0.0
Youth policy	0	0	7 700	0.0	0.0	0.6
Social assistance	201	1 893	5 719	0.3	0.5	0.4
Government[b]	855	12 508	40 099	1.1	3.5	2.9
Law-enforcement organs	0	0	4 707	0.0	0.0	0.3
Other	268	4 433	1 588	0.3	1.3	0.1
Chernobyl related	0	0	56	0.0	0.0	0.0
Compensation	0	0	53 563	0.0	0.0	3.9
Reimbursement for school uniforms	2 617	35 035	0	3.4	9.9	0.0
Budgetary loans	0	0	5 000	0.0	0.0	0.4
Settlements with higher budgets	37 400	80	0	48.5	0.0	
Total expenditures	77 138	352 849	1 374 700	100.0	100.0	100.0

[a] 1993 includes 8958 for fund for unforeseen expenditures.
[b] Includes deputy, soviet and administration activity.
Source: Petrodvorets government documents.

the district government as well as a more fiscally responsible local government during the transition period. Table 7.2 and Figure 7.1(b) show that relative spending priorities changed substantially at the local level with the highest proportion of expenditures made locally on education (36.9%), housing and communal economy (23.9%) and then health (21.6%). These trends hold stable and even intensify in the first six months of 1993 with education being 48.9% of budgetary expenditures, health 21.0% and housing and communal economy 19.4%. The relative change between housing and communal economy and health is probably not significant in that the data only show half a year. The increases do not necessarily mean that more funds are allocated to these expenditure items because we do not have the information on expenditures in Petrodvorets by the St Petersburg government; but the focus on expenditures in health, education and housing and communal economy show the formal local priorities. These three areas are areas

of traditional local responsibility stretching back in time to the Tsarist period of Russian history.

The new budgets show substantial continuity with the past in terms of local government functions. There are, however, other less noticeable changes in the formal budgets that show the increased local urban development role through fiscal decentralisation and growing budgetary responsibility at the local level. The formal functions have changed to include social policy, local economic policy and additional responsibilities in the local economy. First, the district increased its expenditure responsibility in the communal economy, making expenditures in 1993 that were made by St Petersburg in the past. Petrodvorets made expenditures for law enforcement, previously a responsibility of St Petersburg and not a local responsibility (Table 7.2). This shows some fiscal decentralisation in the expenditure sphere. Second, the law enforcement expenditures also represent local initiative, because the district created its own local police force in 1992

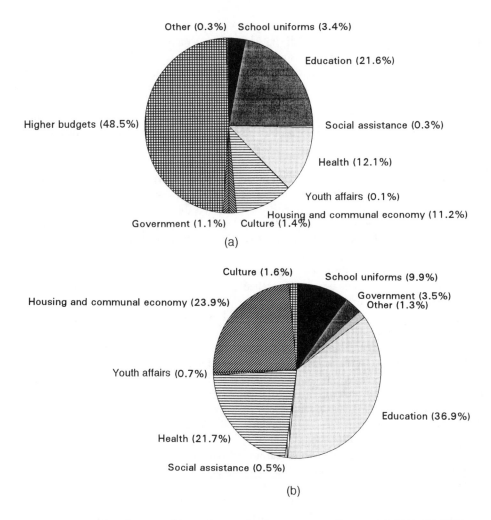

Figure 7.1 *Percentage distribution of Petrodvorets budgetary expenditures: (a) 1991; (b) 1992*

using independent revenue sources. Third, similar budgetary initiative is seen in the inclusion in the formal budget in 1992 of expenditures on the Bureau on Youth Affairs. This bureau was created to function as a type of community centre for the district's youth. This was also created in 1992 on the basis of local initiative using locally accumulated revenue sources. A fourth formal budgetary change occurred in 1993 with wage compensation added to the local budget. Compensation is a category of social welfare where expenditures are made usually on wage compensation for price increases for food and children's clothes. Other levels of government also funded wage compensation. But like Yaroslavl', Petrodvorets attempted to increase compensation and public assistance. A fifth significant

change is the beginning in 1993 to make loans from the local budget to support local economic activity. This was done because of enabling legislation at the level of St Petersburg city. Despite the substantial continuity with the past for the formal functions of local government, significant changes in local governance did begin to occur immediately after the disintegration of the FSU, often at the initiative of district government showing priorities in some new areas of urban development.

Local government initiatives

Local government initiatives are an important area for seeing the changes in local government responsibility in the urban economy during the transition. Local government initiatives can vary greatly across space, in part because of differential ability to form budgetary surplus and extra-budgetary revenues. In the Soviet period, the local government did not independently determine the structure of expenditures from these two sources. The local government did not accumulate investment resources to make these expenditures but would negotiate with higher levels of government and/or enterprises and ministries that functioned on their territory to make these kinds of unplanned expenditures. One official explained that in the past, they would seek out sponsors rather than funding (and fending for) themselves. On one hand, the previous system was easier because it relied solely on negotiation talents while the new system relied on financial expertise and management to accumulate resources. On the other hand, in the previous system independence of action was severely curtailed, while in the new system independence was fostered.

Information on local initiative and legislative priorities is culled from interviews, hundreds of local government resolutions, and a data base constructed from the resolutions. The data base consists of two major categories – political and economic. Each category is then subdivided; political resolutions are categorised into organisational, territorial identity or the environment. Organisational consists mainly of personnel matters, monitoring, elections, and legislative issues including protests and inquiries. Territorial identity includes local awards, borders, monuments, place-name changes and public morale. The economic category is subdivided into agriculture, the budget, industry, organisational, privatisation, social welfare, trade/commerce and urban development. This level of detail gives more specific information on the nature of the local government role in the urban economy than do general budgetary categories.

Already mentioned above are several examples of local government initiative, most of which occurred in 1992. In 1993, the situation changed substantially with respect to local initiative. The local government changed from a proactive stance to a planning one. The fiscal situation of the district was troubled; Petrodvorets was one

of several districts that received subsidies from St Petersburg. The St Petersburg planned budget shows that about half of the districts were to receive budget subsidies in 1993; while local government officials said that by September 1993, Petrodvorets was one of four subsidised districts in St Petersburg. The data base supports the lack of budget surplus in both 1992 and 1993. The topic does not appear in any resolution over the two-year period, although the chairman of the district soviet, Nikolai Ul'novich Marshin, asserted that 100 million roubles from the surplus revenues were spent on the health sector.

For Petrodvorets, initiative was also curtailed using revenues from extra-budgetary sources. The few initiatives funded in 1992 were done so using extra-budgetary revenues. But extra-budgetary funds decreased substantially between 1992 and 1993 because tax fines no longer were an extra-budgetary revenue source. In addition, the soviet reportedly lost its access to its extra-budgetary fund due to "scandals" and conflict with the executive branch. No accounting of 1993 extra-budgetary revenues or expenditures apparently existed or was not available even to workers in the executive branch. Such accounting usually appears in the *malyi* soviet resolutions. With the loss of extra-budgetary funds, the district lost some ability to construct independent expenditure policy for urban development or even to have substantial influence over spending.

Initiative in planning for future economic development is evident in several ways. First, in line with its declaration to be a tourist centre, the district formulated plans to build an international conference hotel and chose two pre-revolutionary buildings as potential sites. Second, both the soviet and the executive contracted with separate consultants to draw up new general plans for land use and urban development. This second planning initiative is indicative of the competition and conflict that has existed between the executive and legislative branches in Petrodvorets. The situation has caused duplication of work and additional expenditure of scarce resources.

The data base information highlights areas of legislative priority and potential initiative for local government. In Petrodvorets, the economy was of greater priority than political issues (Table

7.3). Over 60% of local legislation focused on the economy rather than on political issues. Within the economic realm, local legislation in 1992 was aimed mainly at issues related to urban development (34%), the budget (22%), organisational (20%), social welfare (11%) and privatisation (9%). Urban development legislation focused on infrastructure (including housing and transportation), land use, capital improvements, property transfers, socio-economic planning, mass media and leasing. These are new functions for local government. In the past, land use, property transfer and leasing of property were not local functions but were central functions. They are currently in the local realm due to federal law, but each local government implements that law differently. Mass media and local television programming is a new function at the initiative of the local government. Petrodvorets supports the local newspaper and helped establish a local television channel to service local residents.

Social welfare and privatisation as local legislative priorities are not leftovers from the past. The priority given to social welfare legislation in Petrodvorets firmly places this function as a local one. The initiatives consisted mainly of public assistance measures. Privatisation legislation focused on privatisation of housing and the distribution of privatisation cheques. Local governments are responsible for privatising local property, including housing, according to central legislation.

Legislative priorities changed in Petrodvorets between 1992 and 1993 (Table 7.3). By 1993 shaping the urban environment became a local priority. Increased priority was given to urban development (46% of economic resolutions) and to social welfare (16% of economic resolutions). Significant relative declines were seen in attention given to organisational matters (from 20 to 9% in 1992 and 1993 respectively) and to privatisation (from 9 to 4% in 1992 and 1993 respectively). This is due to a steep decline in the monitoring of economic decisions, from 42 resolutions in 1992 to only 4 in 1993 and in the attention given to the privatisation of housing. The decline in attention to privatisation, however, is countered by an increase in the attention given to leasing and property transfer in the urban development category. Leasing and property transfers are two ways the local government can directly influence land use and economic activity. The increased attention to social welfare is mainly due to additional legislative attention given to health matters; this

Table 7.3 *Data base accounting for Petrodvorets*

Category	1992		1993	
	Number	Percentage	Number	Percentage
Economic	283	100	180	100
Agriculture	5	2	1	1
Budget	62	22	40	22
Industry	3	1	0	0
Organisational	57	20	16	9
Privatisation	25	9	8	4
Social welfare	30	11	28	16
Trade/commerce	7	2	5	3
Urban development	95	34	83	46
Political	162	100	109	100
Environment	0	0	0	0
Organisational	151	93	99	91
Territorial identity	11	7	10	9
Unknown	2		4	
Total	448		293	

Source: Petrodvorets *malyi* soviet resolutions.

supports the information given by the district soviet chairman that health was a major priority of the district soviet. Many local officials spoke of problems with the distribution of medicines.

Petrodvorets has also made some significant initiatives in the political realm. Although the vast majority of political legislation is organisational in nature (i.e. dealing with personnel matters in the government), a small portion does focus on territorial identity (Table 7.3). Territorial identity legislation in Petrodvorets during 1992 was comprised mainly of place-name changes, public celebrations (public morale) and specification of local borders. In 1993 legislation focused on establishing local monuments and on public celebrations. Also during this period, the local soviet set legislation in motion at the federal level to re-establish the pre-revolutionary name of the town at Peterhof. Local documents refer to the district now interchangeably as Petrodvorets, Peterhof, the district and the town (*gorod*). By making territorial identity a local priority, the local government establishes its influence over activity within the jurisdiction and engages in place advertisement.

Intergovernmental relations

During the Tsarist era, cataclysmic social events, such as war and revolution, and major social restructuring, such as the emancipation of the serfs, served as catalysts to the restructuring of intergovernmental relations. In the post-Soviet period, the disintegration of the Soviet government and economic crises have contributed to the restructuring of central–local relations. Intergovernmental relations have clearly undergone restructuring during the transition period, in part due to fiscal decentralisation and the decentralisation of other responsibilities discussed above.

The structural separation of local government from higher levels of government and local government from the Communist Party immediately after the disintegration was clearly a turning point for post-Soviet local government. The local function of administering central directives, characteristic of both the Tsarist and Soviet periods, is still theoretically in place such as is the case

with mandated local privatisation. But the conflict between branches of government and levels of government that culminated in the bloody events at the Russian Parliament in autumn 1993 underscore the reduced ability of the centre to conduct local affairs. The conflict between the St Petersburg government and the district governments as the city government attempted to reduce district level government is also indicative of increased local strength during the transition period.

In Petrodvorets, conflict between the St Petersburg government and the district soviet is somewhat minimised due to the financial subsidisation that Petrodvorets requires. However, some instances do exist. The city of St Petersburg ignored the independence declaration by Petrodvorets. Some legal protests have been lodged between the district and city governments due to resolutions concerned with the management of the local economy, an area of conflict present in both the Tsarist and Soviet periods as well. Conflict between the executive and legislative branches of government in Petrodvorets is more prevalent than conflict with St Petersburg government. For example, in 1992, the local soviet had to take seven votes before an administrator could be agreed upon. Also, the competitive nature of the two branches is seen in the construction of separate development plans mentioned above.

In Petrodvorets, as in St Petersburg as a whole, an unusual mixture of decentralisation and recentralisation has been occurring. In 1992, after the disintegration of the Soviet Union, fiscal decentralisation began as well as the distribution of property among government levels. During this time, a tax collection office was established at the district level to facilitate the collection of city and district taxes and fees. The St Petersburg government decentralised to Petrodvorets district the electrical network and received in return from Petrodvorets three parks to maintain. In a financial sense, this trade is not equal. The maintenance of the electrical network is much greater than the parks; the substantially increased local responsibility for urban infrastructure was thus accompanied by increased expenditure requirements.

Recentralisation at the urban level in Petrodvorets began in 1993. As of 1 July 1993 St

Petersburg took control of the road network and other aspects of the transportation system. While this relieves the district of some substantial financial commitments, it reduces its ability independently to influence this aspect of urban development and infrastructure. From the perspective of St Petersburg, however, it allows for better co-ordination of the urban transportation system while more firmly establishing its dominance in the urban system several months before abolishing the local soviets.

Conclusions

Several aspects of local government functions, initiatives in local economic development, and intergovernmental relations remain stable from the Tsarist period through the immediate post-Soviet period. Substantial continuity occurred despite varying political and economic systems. Local government function throughout Russian history has centred on the provision and maintenance of living standards. Both Tsarist and Soviet local governments functioned as managers of central directives and policies, having a minor formal policy-making role. Neither Tsarist nor Soviet local governments had the capability to form industrial development policies or directly influence the location of economic activity. Pre-*zemstvo* Tsarist and Soviet local governments did not have independent tax or expenditure functions. These latter functions severely limited the local government role in shaping the urban or local economic environment. They are areas, however, in which the Russian local governments gained considerably in the immediate post-Soviet era and then began to lose again after the abolition of the Soviets.

Major social and economic restructuring at the central and local levels allowed for the concomitant restructuring in local government function during the latter years of the Tsarist and Soviet periods. Central–local conflict increased over issues related to fiscal policy. The immediate post-Soviet local government is distinguished by growing fiscal independence, growing ability to manage land-use policy, and increasing direct influence over the economic management of local industries. The new local government functions

represent significant decentralisation of authority and responsibility in the system of government. The local emphases on urban development, social welfare and place recognition as independent legislative priorities and initiatives in Petrodvorets contrast with the formal functions of the past. These new priorities increased the local government capacity to construct or improve infrastructure and influence job creation or job loss.

The increased local responsibility for social welfare represents continuity with the *Tsarist* period but a break with the Soviet period. In the Tsarist period, local government in the form of the *zemstvo* and the village community along with the Church was responsible for assisting the poor and supporting social welfare. Although many aspects of social welfare were part of local government function during the Soviet period, such as housing and food supply, in reality Soviet central government ministries and enterprises provided for the social welfare of their workers. Enterprises provided housing, subsidised meals at the enterprise, food supplies and even recreation. During the Soviet period, social welfare was connected to place of work almost more than place of residence.

In the post-Soviet period, economic restructuring at the central level has led to a restructuring of the location of the provision of social welfare. This represents a reallocation of functions between government organisations and is part of the decentralisation process indicative of the transition. The local government has increased its responsibility in the area of social welfare as higher levels of government focus on other issues. This means that in the near future, the well-being of the population will depend to a large degree on the economic priorities of the local government and the ability of the local government to accumulate investment funds and make public expenditures. This assumes that the central government will continue to de-emphasise its commitment to social welfare in favour of dealing with macroeconomic issues, such as inflation, and that local governments will continue to implement central policy with relatively little oversight from the central government. Although economic stabilisation of macroeconomic conditions has welfare consequences, it is unclear whether or not the benefits of stabilisation will accrue to individuals living throughout Russia. Even when economic stabilisation is achieved, new patterns of central–local relations and new areas of local government responsibility will have emerged that change both the formal and the informal ways of governing in Russian cities.

The emergence of local government in Russia during the transition period has depended to a large degree on spontaneous decentralisation as well as formal central policies. Local government initiative, as seen in the case of Petrodvorets, was found through legislated means as well as through the budget worked out between the legislative and executive branches. With the dissolution of the local legislatures and the transfer of many powers to the executive branch of government that occurred during the autumn of 1993, we are likely to see yet another type of local government emerging during the transition period. The new local government will have as its legacy greater initiative in the local economy, increased social welfare function and increased urban development responsibility. The future local government in Russia will be seen as an active player in both social and economic development of cities. The level of authority and responsibility distributed between levels of government will likely be a negotiated phenomenon that varies by region in Russia.

Acknowledgements

The work leading to this chapter was supported from funds provided by the National Council for Soviet and East European Research which, however, is not responsible for the contents or findings of the chapter. The author would like to thank Mike Longan for his cartographic work and Kurt Fangmeier for his research assistance.

Notes

1. See *Sovetskaya Rossiya*, 30 December 1990, FBIS-SOV-91-003, 7 January 1991 and *Ekonomicheskaya Gazeta* No. 11 March 1992 for laws on increased ability for the local level to

tax, and Zakon "Ob obshchikh nachalakh mestnogo samoupravleniya i mestnogo khozyaystva v SSSR" [USSR law "On basic principles of local self-management and local economic activity"], *Izvestiya*, 16 February 1991, p. 2 for additional rights given to the local level.

2. Laws set the formal boundaries for what is included in the budget that the parliament or soviet passed. Additional revenues outside the formal budget accrue to the city in the form of fines and penalties (e.g. for late tax payments or traffic fines). These revenues are called off-budgetary or extra-budgetary and can be used to make expenditures that do not appear in the formal budget that the soviet passes. Off-budgetary revenues and expenditures exist in capitalist democracies as well and are used by governments to side-step their own bureaucracies.

References

Bahry, D. (1987), *Outside Moscow: Power, Politics, and Budgetary Policy in the Soviet Republics*. New York: Columbia University Press.

Bater, J.H. (1980), *The Soviet City, Ideal and Reality*. London: Edward Arnold.

Berkowitz, D. and Mitchneck, B. (1992), "Fiscal decentralization in the Soviet economy", *Comparative Economic Studies*.

Cattell, D.T. (1968), *Leningrad: a Case Study of Soviet Urban Government*. New York: Praeger.

Fallows, T. (1982), "The zemstvo and the bureaucracy, 1890–1904", in Emmons, T. and Vucinich, W.S. (eds), *The Zemstvo in Russia, an Experiment in Local Self-government*, Cambridge: Cambridge University Press, pp. 177–241.

French, R.A. and Hamilton, F.E.I. (eds) (1979), *The Socialist City*. New York: Wiley.

Hahn, J.W. (1988), *Soviet Grassroots: Citizen Participation in Local Soviet Government*. Princeton: Princeton University Press.

Hough, J. (1969), *Soviet Local Prefects: the Role of Local Party Organs in Industrial Decision-making*. Cambridge, Mass.: Harvard University Press.

Jacobs, E.M. (1983), "The organizational framework of Soviet local government", in Jacobs, E.M. (ed.), *Soviet Local Politics and Government*, London: George, Allen and Unwin, pp. 3–18.

Manning, R.T. (1982), "The zemstvo and politics, 1864–1914", in Emmons, T. and Vucinich, W.S. (eds), *The Zemstvo in Russia, an Experiment in Local Self-government*, Cambridge: Cambridge University Press, pp. 133–75.

Morton, H.W. and Stuart, R.C. (eds) (1984), *The Contemporary Soviet City*. Armonk, NY: Sharpe.

Miller, M. (1967), *The Economic Development of Russia 1905–1914, with Special Reference to Trade, Industry, and Finance*. London: Frank Cass and Co. Ltd.

Mitchneck, B. (forthcoming 1994), "The changing role of the local budget in Russian cities: the case of Yaroslavl", in Freidgut, T. and Hahn, J. (eds), *Local Power and post-Soviet Politics*, Armonk, NY: Sharpe.

Mitchneck, B. (April 1995), "An assessment of the growing local economic development function of local authorities in Russia", *Economic Geography*, **71**.

Mitchneck, B. and Plane, D. (1994), "Migration and the quasi-labor market in Russia", paper presented at the 34th European Conference of the Regional Science Association, Groningen, the Netherlands.

Petrov, N.V., Mikheyev, S.S. and Smirnyagin, L.V. (1993), "Regional differences in the Russian Federation: social tensions and quality of life", *Post-Soviet Geography*, **34**: 52–66.

Ross, C. (1987), *Local Government in the Soviet Union: Problems of Implementation and Control*. New York: St Martin's Press.

Ruble, B.A. (1991), *Leningrad: Shaping a Soviet City*. Berkeley: University of California Press.

Shapovalov, V. (1984), *The Moscow Soviet*. Moscow: Progress Publishers.

Starr, S.F. (1972), *Decentralization and self-government in Russia, 1830–1870*. Princeton, NJ: Princeton University Press.

Starr, S.F. (1982), "Local initiative in Russia before the zemstvo", in Emmons, T. and Vucinich, W.S. (eds), *The Zemstvo in Russia, an Experiment in Local Self-government*, Cambridge: Cambridge University Press, pp. 5–30.

Taubman, W. (1973), *Governing Soviet Cities, Bureaucratic Politics and Urban Development in the USSR*. New York: Praeger.

Vinogradoff, P., Sir (1979), *Self-government in Russia*. Westport, Conn.: Hyperion Press.

8
Continuity and change in the post-Soviet countryside

Judith Pallot
University of Oxford, UK

The changing aims of agrarian reform 1985–1995

At the time that the USSR was dissolved at the end of 1991, an agricultural reform programme was already being implemented in the 15 constituent republics of the USSR. It was one of the centre-pieces of Gorbachev's *Perestroika* strategy. Gorbachev understood that failures in the agricultural sector were putting a break upon the USSR's economic development and, in particular, he wanted to reduce the country's dependence upon imported food and feedstuffs which were absorbing too great a portion of the country's foreign currency earnings. Gorbachev was not the first Soviet leader with this goal but he differed from his predecessors in the approach he adopted to realising it. Rather than increase the already substantial investments going into agriculture (in the final years of the Brezhnev regime approximately one-quarter of all new investment in the Soviet economy was directed towards agriculture), he committed the country to a course of radical agrarian reform and rural renewal. Through introducing organisational changes at farm level and by upgrading rural living standards, Gorbachev hoped to improve farm efficiency and stabilise the rural labour force which, in the traditional farming regions, had been depleted qualitatively and quantitatively by selective out-migration (Rybakovsky and Tarasova, 1989, pp. 70–1). The innovation in Gorbachev's reform was that it encouraged the development of individualised farming which had been anathema to a socialist agricultural system since collectivisation in the 1930s. Measures were introduced which enabled individual or small groups of collective and state farm workers to contract to work a parcel of land or to provide a specific service for their parent farm for a fixed period. The contracts, termed *arendniye* or *semeniye podryady* (rental or family contracts), represented a substantial devolution of responsibility within the organisational structure of collective and state farms (Pallot, 1991, p. 91).

The contract system was introduced under the banner of restoring the traditions and habits of the "peasant farmer" to Soviet agriculture. Over-centralised management and the insistence on collectivised work patterns, it was argued, had separated farm-workers from the land with negative consequences for efficiency and productivity. The contract system was designed to re-establish the farmer's close relationship with the land. This was a proposal that resonated strongly with the Soviet population. Rural nostalgia and fears about the loss of "Russian" culture, symbolised by the abandonment of villages in central and northern Russia, had taken hold in the previous two decades and Gorbachev's measures seemed to respond to these anxieties. It was perhaps inevitable that, as events accelerated towards the final crisis in the communist system in the second half of the 1980s, the agrarian reform began to take a more radical turn. If the solution to the USSR's agricultural crisis lay in developing more independent forms of labour organisation, it was argued, why not go the whole course and allow private farms to be set up in the Soviet countryside? By the time Gorbachev stepped down from power, legislation was on the statute books which enabled collective and state farm-workers to leave their *kolkhozy* and *sovkhozy*, taking with them a parcel of land on which to set up their own farm. There were parallel provisions for non-agricultural workers to "return" to the countryside to set up private farms (Pallot, 1990).

The independent farms set up under the provisions of Gorbachev's reform legislation did introduce a limited form of private enterprise into the Soviet countryside. Rather like personal subsidiary farming (*lichnoye podsobnoye khozayaistvo*) practised by collective and state farm-workers and until the 1980s the principal form of private economic activity permitted in Soviet agriculture, independent farming was hedged round with various conditions to prevent its development into a form of capitalism. Gorbachev was careful to present his reform as wholly consistent with a communist path of agrarian development. The new farms were described in official publications as labour-production farms (*trudoviye khozyaistva*), a

Geography and Transition in the Post-Soviet Republics. Edited by M.J. Bradshaw.
© 1997 M.J. Bradshaw & Contributors. Published 1997 by John Wiley & Sons Ltd.

specific type of non-acquisitive peasant farm identified by the rural sociologist, A.N. Chayanov, in the 1920s, and restrictions were placed upon the hiring of labour, on property rights and on farm size. There could be no hint under Gorbachev that the abandonment of collectivised farming was in prospect. The independent farm sector was expected to make a contribution to overall output but the main role it was supposed to play in the USSR's agricultural revival was to spur the socialised sector on to greater efficiency and productivity through competition. In other words, the development of independent farms was intended to stimulate *perestroika* of the *kolkhozy* and *sovkhozy*.

At the time of the USSR's collapse the number of independent farms that had been formed stood at 49 000 (Wegren, 1993, p. 50). Their average size was 41–42 ha. These farms were unevenly distributed between the republics and regions of the USSR. The greatest concentrations were to be found in the non-black earth zone in European Russia and in the USSR's mountainous peripheries in the Transcaucasus, the Carpathians and Altai. Relatively little headway had been made in the principal cereal-farming regions in southern Russia, Ukraine, west Siberia and northern Kazakhstan, or in the industrial cotton-producing regions of Central Asia. Wherever land was at a premium, the response to the reform was limited (Wegren, 1993, p. 52). Furthermore, its appeal seemed greater to non-agricultural workers than to people already in the sector. The initial results of the reform were thus disappointing. Nevertheless, every region and republic of the former USSR had its complement of independent "peasant" farms by 1 January 1992.

In developing their own agrarian policies the leaders of the post-Soviet successor states have mostly adopted a cautious approach. However, there are notable exceptions. In the Baltic states, even before independence was formally granted, a free farm movement had developed and new governments came to power partly on the promise of an immediate restoration of land to pre-occupation owners. In Estonia, for example, the 1989 Farm Law guaranteed the restitution of farms to anyone who could prove that their family was the owner in 1939. By 1990 some 4000 independent farms had been formed, although

since independence formal registration of restored farms has proceeded rather slowly (Unwin, 1994a). In the Transcaucasus, radical land reform was also a feature of the immediate post-independence period. In Armenia land reform legislation in 1991 effectively divided the republic's agricultural land between farm-workers to create over 200 000 small farms with an average size of under 1 ha. There were specific historical and geographical reasons why land reform should be pursued vigorously in the Baltic states and the Transcaucasus. In the former, "late" collectivisation meant there were people still alive in the 1990s who had farmed independently or who could recall where their patrimonial lands lay. Despite industrialisation during the Soviet period and deportations, the Balts had retained a strong sense of their "peasant heritage". In the Transcaucasus there was also a strong tradition of private farming; Georgian and Armenian farmers had profitably supplied the northern industrial markets with subtropical fruits and vegetables grown on their "private plots" during the communist period. Furthermore, in the Caucasus mountains collective farms had never taken a firm hold and long before the USSR's collapse farming had effectively reverted to an individual basis.

Elsewhere in the former Soviet Union, where there were no such special factors pushing in the direction of individualisation, there were good reasons why post-Soviet governments should proceed slowly. These included fear of disruption to the food supply, conservative opposition to radical reform and an apparent preference for the status quo on the part of a majority of farm-workers. While not necessarily acknowledging its author, most post-Soviet leaders have simply built upon Gorbachev's reform and, with greater or lesser degrees of enthusiasm, have promoted the formation of individual farms through a process of voluntary cessation from *kolkhozy* and *sovkhozy*. In most of the successor states, the rights of the new owners over their farms are hedged with restrictions. In Turkmenistan, for example, there is an upper limit on the size of the farm any individual can "own", with further limits on the amount of additional land that can be rented. In Central Asia, Kazakhstan included, and in Azerbaijan, the state has retained ownership of the land. Individuals setting up private farms can be granted a 99-year leasehold, with the right to bequeath it to heirs, but they do not own the land. In the case of Kazakhstan and Kyrgyzstan the ban on private ownership is clearly motivated by fears of land passing to non-indigenous groups in the population. Even in those states in which private ownership has been permitted, such as in the Russian Federation, there are restrictions on the disposal and use of land. Most states require anyone taking up a private farm to bring land into productive use within a set time period and everywhere there are prohibitions on non-nationals investing directly in agricultural land (CDPSP, 1994 [19], p. 13).

It is difficult to find reliable figures of the number and geographical distribution of independent farms that have been formed in all the successor states since the collapse of the USSR, although for the Russian Federation, Craumer has analysed the distribution of private farms in 74 *oblasts* in relation to a number of socio-economic variables (Craumer, 1994). Map 8.1

shows the absolute number in the former Soviet Union in mid-1993, updated where figures are available (SWB, 1992 SU/WO24 A/9 [54]). The share of agricultural land occupied by independent farms remains small except in the Baltic states and Transcaucasus. In two states, Latvia and Armenia, the pursuit of policies to divide collective and state farmland immediately after independence has resulted in major landscape change, large farms being replaced by a multitude of smallholdings a handful of hectares in size. Subdivision has also proceeded rapidly in Georgia, although here the process of agricultural restructuring has been complicated by civil war and economic collapse. Figure 8.1, overleaf, shows the subdivision into micro-farms of a former state farm in the republic. Elsewhere landscape change has been less profound. In the Russian Federation the share of farmland owned by individual farmers averaged under 5% at the end of 1993 with only a small number of *oblasts* with more than 10% (Craumer, 1994, pp. 332–3; *Ekonomicheskoye polozheniye*, 1994, p. 146). By

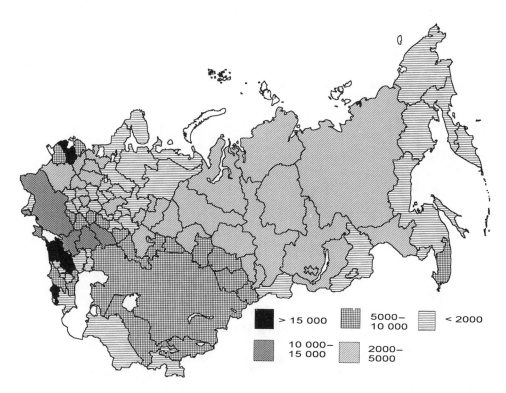

■	> 15 000	▓	5000–10 000	▤	< 2000
▨	10 000–15 000	▥	2000–5000		

Map 8.1 *Number of independent farmers in the FSU, mid-1993*

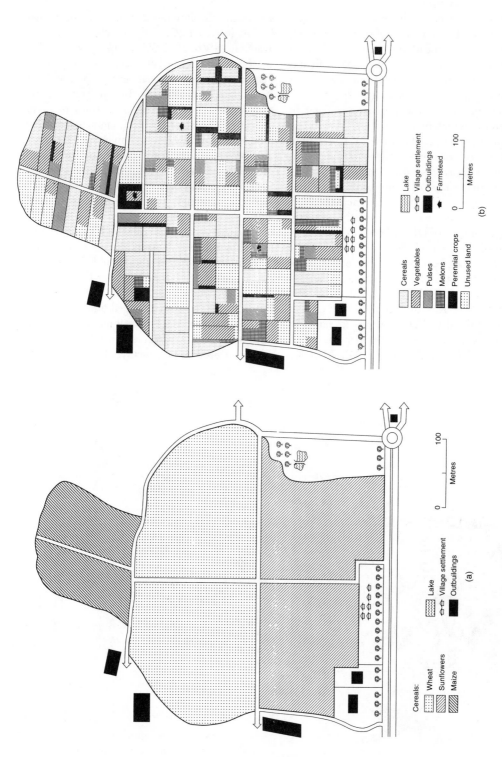

Figure 8.1 *The landscape of decollectivisation: the layout of Chocti village in eastern Georgia (a) before and (b) after the early 1990s*

Map 8.2 *Share of arable land in the Russian Federation owned by independent farmers by oblast, 1 January 1994*

the end of 1995 some 283 000 private farms had been formed in the Russian Federation but the annual increment of new farms has slowed down in recent years and, furthermore, approximately 9–10% have been going out of business annually. Map 8.2 shows how this land is distributed between *oblasts*. Regional patterns of private farm formation are complex and defy simple explanation (Craumer, 1994, p. 348). In contrast to how things stood in 1990, the greatest development of independent farms is now to be found in the principal cereal-farming regions in the south of the country where agricultural resources are best. This, no doubt, reflects the strengthening of individuals' rights *vis-à-vis* their *kolkhozy* since Yeltsin came to power (Wegren, 1995, p. 177). The same concentration of independent farms on the best agricultural land is also to be found in Estonia (Unwin, 1994b).

The promotion of individual farming in the post-Soviet states has been pursued within the context of differing attitudes to the pre-existing socialised farms. Roughly classified, the successor states fall into three groups: those in which measures have been introduced aimed at abolishing all vestiges of the previous farm structure; those, at the other extreme, in which the existing structure of collective and state farms has been left more or less intact; and, finally, those in which measures have been introduced designed to transform collective and state farms into more efficient, "Western type", large farms. Assigning the successor states to these groups is complicated by shifts in policy that have taken place since 1990–91. The Russian Federation, for example, initially belonged in the first group but since 1993 it has abandoned early intentions of the immediate dissolution of collective and state farms. There has been a general tendency for

115

states hostile to the principle of collective and state farming to moderate their policies towards large farms in the five years since independence. There has been greater consistency in policy among the group of states that have continued their support of collective and state farms. In Uzbekistan, the state has maintained a firm control over agriculture and apart from divesting itself of responsibility for the country's least efficient *sovkhozy*, the post-communist authorities have done little to upset the agrarian structure they inherited from the Soviet period. In Moldova there is a clear reluctance to introduce any measures that would undermine the country's agro-industrial complex which prior to the USSR's collapse had been remodelled. Moldova's economy was based upon supplying the Soviet market with processed fruit and vegetables and wine. The Agrarian Democratic Party which represents the interests of the large-farm managers and rural officials, has won a majority in every post-Soviet election in Moldova and it continues to exert a break on change in the agricultural sector. In the majority of successor states a convergence has taken place in attitudes towards the third group described above. Thus, policies to transform collective and state farms into more efficient, but still large-scale, enterprises had by 1995 been adopted by most states. But they have been pursued with differing degrees of enthusiasm. While in the Russian Federation, President Yeltsin has consistently argued for a fundamental change in the organisational structure of the country's large farms, in both of the other Slavic states, Ukraine and Belarus, little more than lip service has been paid to the need.

The principal agent of change chosen by the post-communist states to effect a transformation of *kolkhozy* and *sovkhozy* is "privatisation" by which is meant the transfer of farm ownership to private citizens. As in industry, privatisation in agriculture has generally involved allocating shares to the workforce, including to pensioners retired from farm work but still resident on farms. Wherever such privatisation is taking place, collective and state farms are being redefined as "joint-stock companies", "limited liability companies", "associations" and "co-operatives". But the "privatisation" of collective and state farms in their pre-existing form is also allowed for in some states (in

Belarus, for example, a 1993 law recognised *kolkhozy* and *sovkhozy* as private landowners). It would, in fact, be more accurate to describe the privatisation of large farms in the successor states as "destatisation" since the process involved is essentially one of cutting farms loose from the state. The belief is that, left to their own devices and without the protection and support the communist state gave them, the *kolkhozy* and *sovkhozy* will either become more efficient (by divesting themselves of surplus labour, streamlining management and being more responsive to the market) or they will fail and fall apart. While there is evidence that collective and state farms are genuinely being transformed in some places, as in Estonia where far-reaching changes have been introduced into the economy, in the majority of successor states the retreat of the state from agriculture has been partial at best. The system of grain procurements was still in operation in Ukraine at the end of 1994, despite the insistence of the World Bank that it be dismantled. The development of a market in agriculture in all the successor states has been further stifled by the continuation of substantial subsidies to the agricultural sector, the introduction of import tariffs on basic foodstuffs and the failure of enterprises upstream and downstream of the primary agricultural producers to be privatised. These circumstances, together with the fact that land conveyance is subject to restrictions, means that it is difficult to predict how far-reaching the reform of collective and state farms is likely to proceed in any of the successor states. To date, the evidence is that the changes have been mainly cosmetic. The Russian Federation has not escaped this criticism.

Yeltsin's intention of dissolving collective and state farms was announced over the winter of 1991–92. Legislation approved by the new Russian parliament required all collective and state farms in the Russian Federation to set up internal land reform committees whose task it was to draw up plans for farm privatisation (Wegren, 1993, 1994a). Within three months the new status of each collective or state farm was to be registered with local authorities and by the end of the year reorganisation in line with the new status was to have taken place. A different fate awaited the 1000 most indebted farms in the country; these were to be dissolved and their land and working capital divided between members as independent farms. These measures left no room for doubt that the intention of the new regime was to deliver a terminal blow to the collective and state farm system. The shape of the agrarian landscape that was to emerge from the ruins of the old was not clearly spelled out in the reform legislation. A variety of new farm types was allowed for, but the expectation was that, in time, market competition would produce an appropriate mix of large and small farms and of different management types. The priority was to remove the legal protection collective and state farms had previously enjoyed and thus leave them vulnerable to the eroding effects of the market. The reform strengthened the mechanisms for individual departures, which were expected to undermine farms "from within" through a constant haemorrhage of land and of the most enterprising farm-workers. The reform's results did not live up to these expectations. The number of individual farms formed did increase rapidly in the year following the promulgation of the reform legislation, but collective and state farms proved themselves able to withstand change. By the middle of 1991, Yeltsin was forced to compromise on some of his measures (Uzun, 1993). Notably, land reform committees were allowed to recommend that their farm retain its present status. Planned mass bankruptcies of the most loss-making farms did not take place. More telling for the prospects for a radical restructuring was that it soon became clear that many farms registering a change of status to become a joint stock company or peasant co-operative, had been engaged in little more than an exercise in "changing nameplates" (Pallot, 1993). Organisationally, they remained much the same as before.

Yeltsin's response to the evident ability of farms to accommodate legislative changes without this affecting their *modus operandi*, was to issue further decrees. The most important was the Land Decree of October 1993 which provided for all farm members, including pensioners, to be issued with title deeds corresponding to their share of their farm's land and working capital (CDPSP, 1993 [44], pp. 8–10; van Atta, 1993; Wegren, 1994a). Simultaneously, the government, in the wake of the October 1993 storming of the White House still united behind the idea of

radical reform, gave its official endorsement to a method of farm subdivision involving the auctioning of farm shares which had been developed as a joint pilot project between the Nizhny-Novgorod government and the International Finance Corporation (CDPSP, 1993 [44], p. 10). These measures fared little better than the previous ones. By the beginning of 1995 there were few commentators who were prepared to vouch that the changed legal status of the country's collective and state farms had effected a fundamental change in agricultural organisation. As Yegor Gaidar' noted, the reform had ground to a halt (CDPSP, 1994 [23], p. 2). This was also admitted by Yeltsin when, in a state of the nation address at the beginning of 1995, he blamed Russia's agricultural crisis on the failure of land reform. He urged the nation to make 1995 a year of real agrarian reform, reaffirming its aim: "One's own farm on one's own land – that is the reform's long-term goal" (CDPSP, 1995 [7], p. 2).

To summarise, the agrarian reforms pursued by post-Soviet successor states have, on the whole, been limited in scope; individual farming has been promoted within the context of the preservation of large farms which are the direct heirs, and sometimes indistinguishable from, the collective and state farms of the Soviet period. In the successor states which have a conservative approach to economic reform, these policies have been pursued more or less consistently since the collapse of the USSR but in those states committed to radical economic change, current agrarian reform represents a retreat from former more ambitious policies of farm restructuring. In all the successor states political and ideological considerations have figured prominently in the development of policy towards farm structure whether the aim has been to destroy collective and state farms or to retain them. No doubt because of this, agrarian reform has been developed in most of the successor states more or less independently of agricultural policy.

Inertia in the post-Soviet countryside

Map 8.3 shows the share of agricultural land in the Russian Federation owned by collective and state farms and their post-Soviet successors, the joint-stock companies, co-operatives and associations. Although some inroads have been made since 1991, large farms remain the dominant feature of the Russian countryside. It is difficult to escape the conclusion that the Russian countryside is in the grips of powerful forces of inertia. Contrary to what was thought in the early years of the agrarian reform, resistance to change is widespread among ordinary members of the farm population and is not just a characteristic of farm managers whose reluctance to surrender power is well understood. One of the ironies of the past decade is that agrarian reforms launched under the banner of returning land to the peasants, has floundered because those for whom the reforms were intended have been behaving in the most "peasant-type" of ways – they have been playing safe and avoiding risks (Nikol'skii, 1993).

At the level of *kolkhozy* and *sovkhozy* and of their successor large farms the pursuit of survival strategies has resulted in the maintenance of the diversified patterns of production that were characteristic of farms during the communist period. Thus, while agrarian specialists had predicted that the substitution of the market for the plan in agriculture would result in the emergence of greater enterprise and regional specialisms in agriculture, these expectations have not been realised. The first priority of farms is to meet their own and their region's requirements for food and feed. Low agricultural prices and delays in payment have reinforced this tendency. Output of agricultural products has fallen everywhere in the former communist states but this has been accompanied by a fall also in the proportion of produce marketed. In Ukraine, previously a principal grain-exporting region of the USSR, it has been calculated that some 60% of all cereal output is currently being retained on farms. Shortages of machinery, in part the consequence of

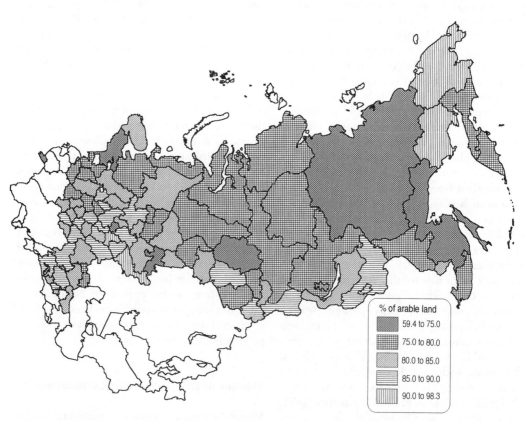

	% of arable land
	59.4 to 75.0
	75.0 to 80.0
	80.0 to 85.0
	85.0 to 90.0
	90.0 to 98.3

Map 8.3 *Share of arable in the Russian Federation under collective and state farms, joint-stock and limited liability companies and other large-scale enterprises, by oblast 1 January 1994*

the unfavourable terms of trade between towns and countryside, constitute a major obstacle in the way of farms pursuing internal reform strategies as any new production units would be prohibitively expensive to equip. With approximately half the tractors, combine harvesters and ploughs on farms in the Russian Federation out of order, the prospects for successful farm divisions must be remote (SWB, SUW/0367 WC/3, 20 January).

If the economic environment is not conducive to large farms striking out in new directions, the same applies several times over at the individual level. Collective and state farm members, it seems, are extremely reluctant to pursue any course of action that would upset the traditional relationship they have with their parent farm because the latter gives them some degree of security and protection from the uncertainties of the economic transition. This must be a major explanation for the failure of individuals to take up independent farming despite the opportunities escape from the *kolkhozy* would give them for material gain and personal freedom. It also explains why farm members have been content to allow the power structures associated with the socialised farming system to be carried over to the "new organisational forms". Political inertia at the local level, in turn, has allowed conservative agrarian interests to remain dominant in national politics. In the Russian Federation, the agrarian lobby, consisting of the old collective and state farm "bosses", has successfully argued for subsidies for the agricultural sector which, in turn, have helped large farms survive in an otherwise hostile economic environment (Konovalov, 1993; Yakovleva, 1993a).

The reluctance of individual farm members to strike out on their own and leave the security offered by their parent farm is understandable, given the difficulties they are likely to encounter. These have been well documented in the Russian press in the past decade. Small farmers complain of a lack of state support for their efforts (CDPSP, 1994 [34], p. 19 and [36], p. 19). Under Gorbachev, a principal problem was that the Soviet machine industry was not geared to produce the types of technologies needed by small farmers. Today the problem is less an absolute shortage of appropriate technologies than their

price. Inflation and the unfavourable terms of trade for agriculture have meant that few independent farmers can afford to equip their farms properly (Prosterman *et al.*, 1993, p. 32). The problem of the price of inputs is further compounded for independent farms by poor rural infrastructure, weakly developed marketing structures and the unavailability of cheap credit which, if secured, is quickly eroded by inflation. On the social and psychological level, independent farmers have often had to face the hostility of their neighbours and withdrawal of access to vital rural services which are still the monopoly of large farms. The location of private farms, often far from the main highways, can further serve to underline such farmers' isolation. A comparative study of Rostov and Kostroma *oblasts* revealed that the location pattern of private farms differs according to local political and economic factors; some local administrations are more sympathetic towards the independent sector than others, and this can affect the relative accessibility to communications and other services of the land made available for the new farms (Wegren, 1994b, pp. 473–6).

The difficulties independent farmers face in the current economic environment is reflected in the generally poor performance of the independent farm sector. Accounting for 4.6% of arable land, independent farms contributed only 1.9% of total agricultural output in 1994 (*Ekonomicheskoye polozheniye*, 1994, p. 156) (Table 8.1). The total number of farms that have gone out of business since 1991 is not available, but for 1995 the number was 19 000 (CDPSP, 1994 [13], pp. 18–19; [49], p. 20). Supporters of radical reform in agriculture argue that the loss of independent farms is merely part of the "normal" market process of "sorting the wheat from the chaff" (CDPSP, 1994 [46], p. 25). It is an argument that cuts little ice with the independent farmers themselves or with those contemplating striking out on their own. Quite reasonably, the latter conclude that it is better to wait to make a move until the economic environment is more favourable for independent farming.

However, there is more to the apparent conservatism of farm-workers than their fear of economic failure; there are positive benefits attached to remaining members of a larger

Table 8.1 *Contribution of different farm types to total agricultural output in the Russian Federation by region (in percentages), 1993*

	Large-scale farms		Independent farms		Private plots	
	% share arable	% share output	% share arable	% share output	% share arable	% share output
North	80.2	70.0	2.4	1.0	11.0	29.0
North-west	76.7	60.0	3.8	1.2	12.8	39.0
Centre	82.1	62.0	3.5	1.3	6.3	37.0
Volga-Vyatka	85.5	59.0	1.8	0.6	5.1	40.0
Central black-earth	84.2	63.0	4.1	1.5	6.6	35.0
Volga	85.0	62.0	6.8	2.6	1.4	35.0
North Caucasus	86.1	66.0	4.2	2.7	2.9	31.0
Ural	82.7	60.0	4.5	1.6	3.0	38.0
West Siberia	82.0	64.0	4.8	2.5	1.6	33.0
East Siberia	83.0	60.0	3.5	1.4	3.3	39.0
Far East	77.5	61.0	6.7	3.9	5.4	35.0
Kaliningrad	74.9	60.0	4.8	1.4	4.1	39.0
Total	83.3	63.0	4.6	1.9	3.4	35.0

enterprise and for supporting the status quo. Collective and state farms have traditionally provided more than just employment for rural denizens. They had been associated with a distinctive "way of life". Albeit sometimes with difficulty, collective and state farms met a variety of their members' social, psychological and welfare needs, including providing them with shops, clubs, entertainment, schooling, health care and housing. An Agrarian Institute poll in the early 1990s showed *kolkhozy* and *sovkhozy* to be responsible for partially covering 80% of the education expenses and 70% of the medical care expenses of the rural population and responsible for meeting 74% of rural housing need (Prosterman *et al.*, 1993, p. 38). The prospect of local authorities assuming full responsibility for these functions is extremely remote at present, so it is not in farm members' interests to act in ways that would undermine the traditional providers.

An even more valuable asset collective farms provide for their members is the land and the material resources to enable the latter to engage in personal subsidiary farming (*lichnoye podsobnoye khozayaistvo*) or, as it is known in Western literature, "private plot agriculture". Private plot agriculture was a very specific form of individualised farming which developed in conjunction with the socialisation of farming in the Soviet Union and East European communist countries.

Indeed, it can be said that the characteristic feature of the communist farming system was its dualism; its combination of large-scale mechanised production using collectivised labour resources with small-scale semi-subsistence production on individual farm members' plots. In the 1930s private plots were a lifeline for collective farm families whose members were poorly remunerated for the work they did in the socialised fields. Although the disposition of land for private plots lay with the parent farm and ownership rested with the state, collective farm-workers became accustomed to viewing their plots as their own. Normally the plots were no more than 0.6 ha in size and typically they were used for the cultivation of vegetable and fruit crops and to support small livestock. During the decades after collectivisation, private plot and socialised farming became inextricably linked. An informal contract developed between farm managements and farm-workers; the former permitted their members the use of collective farm resources for private plot production and, in return, farm-workers consented to work in the socialised fields (CDPSP, 1994 [6], pp. 10–11). The relationship was mutually beneficial. The peasants were able to do well on their private plots meeting their domestic needs and producing surpluses to sell on the market, and the collective farms were, other things permitting, able to meet their centrally determined sales quotas.

During the last decades of the Soviet regime, rising rural wages and changing expectations began to undermine the private sector. The evidence was that the younger generation of agricultural workers was more interested in the pursuit of leisure activities than in working on the household allotments late into the evening. Nevertheless, for a majority of collective farmworkers, the economic stagnation of the final years of the Soviet regime meant the private plot was an important, and reliable, source of subsistence and money income. With the collapse of communism and the removal of controls that used to exist upon private plot agriculture, such as on the number of livestock that could be kept and the size of plot, the sector has been given a new lease of life. In fieldwork conducted in Saratov *oblast* in 1992 a joint American/Russian research team found one state farm on which 20% of the workers had asked, and been granted, an additional 300 m² of arable land to add to their existing private plots (Prosterman *et al.*, 1993, p. 35). Nationally, there was a 15% increase in the area of agricultural land devoted to private plots between 1991 and 1992 and a further increase of 58% by the end of 1993 (Prosterman *et al.*, 1993, pp. 35–6). The change in the conditions for pursuing private plot production since communism's collapse has been described thus by one agronomist: "Whereas before . . . the peasant was always running up against the authorities . . . now everything has been decided in his favour. If someone wants to have two or three cows (previously only one was permitted), why not? Or a flock of sheep, or a gaggle of geese or rabbits or minks. All these are possible as well. Today we can find farm workers with three or four cows, a pair of oxen, a sow and a whole yard-full of geese" (Timofeev, 1993).

The transition period has clearly created conditions which favour the expansion of semi-subsistence forms of farming and which have deepened the peasants' commitment to private plot agriculture. It is a commitment which runs counter to government's attempts to reform Russia's agrarian structure since it strengthens the peasants' ties with their parent farm (CDSPS, 1994 [6], pp. 10–11). As one rural observer has explained, "The owners of these 'micro-farms' have no reason to leave their *kolkhoz*. There is a mutual relationship between peasant farmers and the *kolkhoz* which is based upon cooperation and a rational division of labour . . . so long as the peasant remains a member of his collective farm he does not need to grow his own cereals or cultivate hay. Everything he needs can be provided for by the collective farm, he can simply take (that is, steal) what he needs – grain, cattle cake or hay for his livestock. Yes, and he can even make use of the farm's tractor, car and horses" (Timofeev, 1993). In short, by remaining members of their *kolkhozy* peasants can reap the benefits of individual farming, while avoiding many of its risks (Lerman *et al.*, 1994). It is difficult to predict how in the long term the consolidation of the private plot sector will affect farm structure. In the short term, it is unlikely that the rural population will support any measures that threaten to undermine private plot agriculture. This behaviour is unhelpful to the cause of economic development at the macro level, but it is fully rational from the perspective of the agricultural workers themselves.

"Peasantisation" and "differentiation" in the post-Soviet countryside

The renewed interest of collective and state farmworkers in private plot agriculture is an example of what might be referred to as the "peasantisation" of rural Russia. This term refers to the process whereby people are turning to semi-subsistence forms of farming as an insurance against the risks of the post-communist transition.

Other examples of "peasantisation" include the acquisition of personal plots by non-agricultural workers for subsidiary or "hobby" farming and the fragmentation of collective farmland under the provisions of the land reform. In Armenia and Latvia, large numbers of very small holdings have appeared in the landscape which are primarily engaged in subsistence production (in Armenia the average size is under 1 ha). Elsewhere farms are larger, but many share the subsistence orientation of the Armenian and

Latvian "micro-farms". According to one Ministry of Agriculture publication the vast majority of independent farms created in the Russian Federation since independence belong to the category of "natural producers" engaged in production for family needs (Nikonov, 1994). In one study in Novgorod *oblast* only 16% of newly formed independent farms were classified as "professional farmers" and only a further 9% were rated as having some prospect of developing into professional farmers. The remainder belonged to the category of "marginal farmers" (Petukhova, 1993). One type of marginal farmer that has appeared in the Russian countryside is the "hobby" or part-time farmer. These are urban dwellers who have acquired suburban plots of land, working on them at the weekends and after work. In the period immediately after communism's collapse there was a proliferation in the number of "private" orchards and vegetable plots around Russia's major cities, as land was made available to ordinary citizens under the land reform. As with private plot agriculture, part-time farming is meeting basic subsistence needs of the population. There are in total over 40 million small privately owned plots in the Russian Federation which currently are producing about 35% of the Russian Federation's agricultural output (*Ekonomicheskoye Polozheniye . . .*, 1994, p. 156). Their subsistence function is indicated by the fact that only one-quarter of output currently reaches the market. The rest is consumed by family and friends.

It is likely that in the future regional differences will become obvious in the pattern of "peasantisation" in the Russian Federation. As Map 8.4 shows, part-time and private plot farming is most developed in the northern regions of the Russian Federation in the north and northwest. It is in these regions that there is a tendency for *kolkhozy* and *sovkhozy* to break up and for their land to be distributed into small, marginal farms. This is in contrast to the south where independent farms are formed primarily through the process of withdrawal, are above average in size and have better prospects for developing commercial husbandry. These differences are only to be expected given the inferior agricultural resources of the non-black earth region. In Archangel *oblast*, for example, large numbers of

sovkhozy have simply failed as a result of the withdrawal of state support and farm-workers have taken the land they need to keep their families alive. These, together with independent farms formed as a result of withdrawal at an earlier date, have reverted to a "natural economy" and "declined" to peasant status (Fillipov, 1993). It would seem that a peasantry is in the process of being actively created in the environmentally marginal farming regions of the country. The question for the future is whether such regions will become a fixed feature of post-Soviet Russia or whether they are a transient phenomenon destined to disappear once the economy takes a turn for the better. Global experience would suggest that the former is more likely than the latter.

The creation of a class of marginal or peasant farmers is not the only process reshaping the Russian countryside at the present time. The opposite process, accumulation, is also taking place although it has been dampened by inertia in the collective and state farm sector and the general crisis in agriculture. Nevertheless, some individuals are gaining a disproportionate share of rural economic and political resources and seem destined to continue to acquire in the future. Two examples, taken in turn from the "large-farm" and independent farm sectors, can illustrate the accumulation process in contemporary rural Russia.

The first example is of the subdivision of Burtulino *sovkhoz*, a state farm in Nizhny-Novgorod *oblast*. In response to Yeltsin's 1990–91 decrees, Burtulino state farm transformed itself into a limited liability company. In 1993, the farm was selected for inclusion in the International Finance Corporation and the Nizhny-Novgorod regional government's pilot programme of farm subdivision (IFC, 1993). The *sovkhoz* had 2863 ha of arable land and a total population of 317 of whom 136 were of working age and 181 pensioners. The farm was mixed, with an emphasis on cattle-raising. Under Evgenii Mikheev, who had been appointed as chairman of the *sovkhoz* seven years previously, the farm had become one of the most successful in the district. The aim of the IFC/Nizhny-Novgorod method is to subdivide farms like Burtulino into "viable economic units". It involves members of the farm using "entitlements"

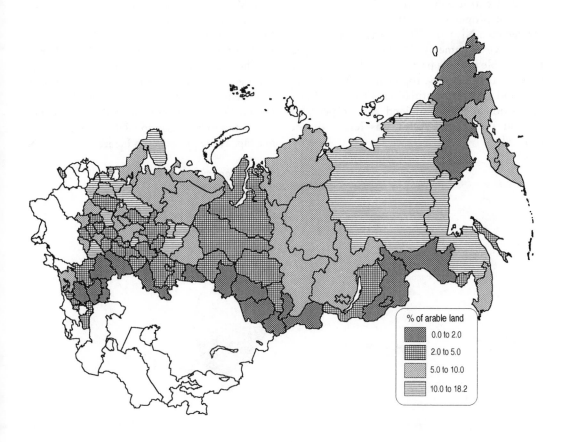

Map 8.4 *Share of arable land in the Russian Federation under the personal subsidiary holdings of farm-workers and urban dwellers, by oblast, 1 January 1994*

to land and fixed capital to bid for lots at an auction. Prior to the auction a committee of experts makes a subdivision of farm resources into a range of possible business units and at the auction partnerships of entitlement holders can bid for these. In the case of Burtulino *sovkhoz*, farm members were each allocated an entitlement of 5.6 ha of land and an average 159 000 roubles worth of property (the latter adjusted in size according to length of service of individual farm members). These were used to bid for lots at the

farm auction in the autumn of 1993. The auction resulted in the farm being divided into two unequal halves (Williams, 1994). One of the new business units took the name of "Niva" and the other "Zarya". At the time of the auction there was, in addition, one individual farm formed and there was one failed bid. The largest of the new enterprises was Niva which formed by a partnership led by the state farm director, Mikheev, who had been able to assemble 226 land entitlements and 56 391 888 roubles worth of capital.

Zarya the smaller partnership had assembled 129 land entitlements and 33 075 819 roubles worth of property. The quantitative asymmetry between Niva and Zarya was matched by a qualitative asymmetry. Mikheev had managed to assemble into his partnership the best of the former farm-workers, together with sufficient land and property entitlements to outbid any opposition for the best land and property of the old state farm. Zarya was left with the residual.

It is clear that in the subdivision of Burtulino farm Mikheev had used the quasi-market set up by the auction to gain command over the best of the former state farm's resources. He did this by persuading the leading workers on the state farm to join his partnership (they became its active partners) and by leasing or purchasing entitlements from pensioners. The Niva partnership had a more favourable ratio of active to non-active partners, and so too to land and capital than the Zarya partnership, as the pie-graph in Figure 8.2 shows. No

Source of land and property entitlements in Zarya partnership bid

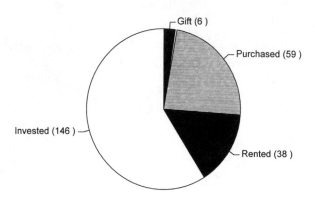

Source of land and property entitlements in Niva partnership bid

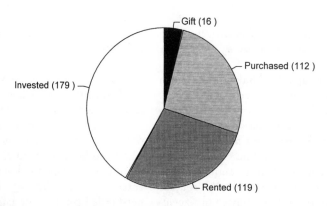

Figure 8.2 *The composition of the Niva and Zarya partnership bids at the auction of Burtulino farm in Nizhy-Novgorod oblast*

doubt, Mikheev's position as farm chairman gave him an advantage over potential competitors in putting together a successful partnership and in winning the co-operation of the farm's pensioners. The terms Mikheev negotiated were extremely favourable to the Niva partnership. For example, pensioners were persuaded to surrender their entitlements in return for payments in kind such as wood deliveries, an annual supply of hay or the promise of some other service. Given the general lack of support for the elderly in rural areas, the services the Mikheev partnership offered did meet a real need, but the fact remains that the partnership was the net gainer from the deals.

The subdivision of Burtulino state farm is an example of how differentiation is taking place in the post-communist countryside. The two farms that resulted from the subdivision were unequal in terms of land, capital and labour resources and in their prospects for development. It would come as no surprise if the Zarya partnership were to fail and the farm fragment, in the process generating peasant smallholdings. Niva's prospects must be brighter. Within a year of the farm's formation, the enterprise had begun to expand. Mikheev built a sausage factory to process the farm's meat products and plans were afoot to expand into brewing. Mikheev's ability to concentrate resources into his hands did not stop at acquiring the best of Burtulino's land, labour and capital. Within the Niva partnership he has emerged as the first among equals, heading a small management team and making himself the sole signatory on cheques.

Although the Burtulino example is in many respects exceptional, the underlying processes that resulted in the division of the farm's resources are a general feature of farm restructuring in Russia. It was the intention in the Nizhny-Novgorod experiment for the subdivision to be "transparent" and "fair" and it was this consideration that lay behind the use of an auction to divide farms. In the event, the auction provided ample opportunity for the state farm management to take advantage of its position in order to ensure it prevailed in the competition. Elsewhere, there are fewer checks against rural élites exploiting their position to achieve this end. "*Nomenklatura* capitalism" is as much a feature of post-communist agriculture as it is of the industrial and the service sector. However, accumulation is not confined to the former rural élites. Judicious "deals" can be used in the independent farm sector as well, as the second example shows.

The example is of a peasant farmer called Ivanov whose "story" was reported in an article in *Izvestiya* in July of 1993 (Suslikov, 1993). Ivanov, a member of a collective farm in Omsk *oblast*, chose to set up an independent farm under the government's land reform legislation on a portion of land separated from his parent collective. He began his career as an independent peasant farmer with two shares of land (equivalent to 24 ha). The farm was small, but it quickly grew to 700 ha through Ivanov's acquisition of land from old-age pensioners and from the local authority's reserve land fund. Ivanov developed a successful wheat business and was able to purchase Western farm machinery. The agreements Ivanov made with pensioners were similar to those in the Burtulino farm subdivision; in return for surrendering their share of land each pensioner received from Ivanov one sack of flour, one sack of sugar, a ton of grain and 20 000 roubles in a one-off payment. Had they understood their economic power these pensioners might have been able to strike a much harder bargain or they might have decided to hang on to their asset. The author of the *Izvestiya* article calculated that Ivanov's pensioners had lost out on 20% of the value of their land. The payment in kind they received was quickly consumed and inflation eroded the value of the money. Meanwhile, Ivanov became richer. Later he joined a co-operative venture with other farmers in the district who had been pursuing similar accumulation strategies. Between them they had amassed the land of 127 pensioners and non-agricultural workers, to a total of 3700 ha.

Future farm structure in the post-Soviet countryside

It is premature to speculate about the future farm structures that will emerge in the post-Soviet countryside. The processes of "inertia", "peasantisation" and "differentiation" that have been described for the Russian Federation are unlikely to be unique to that country, as the other post-Soviet states inherited the same collective and state farm structure and are all now committed to some degree of marketisation in agriculture. However, within and between the countries the outcome of agrarian restructuring will be different, reflecting differences of emphasis in land reforms and in attitudes towards agricultural subsidies, and variations in the successor states' natural and human resource endowments. "Inertia", "peasantisation" and "differentiation" will thus combine differently to produce distinctive agricultural landscapes. One important determinant of the shape of these landscapes will be the extent to which global capitalism succeeds in penetrating the post-Soviet countryside. Currently, foreign capital is being kept at bay by restrictions on land purchase and foreign direct investment. These restrictions could be lifted at some point in the future. Indeed, the suggestion has been mooted in the Russian press of allowing the American McDonald's company to purchase former *kolkhozy* (Yakovleva, 1993b). If such proposals are followed up, the structure of post-communist agriculture will be determined by the global, not national or regional markets, as is presently the case. This would herald a new stage in post-communist agrarian restructuring.

References

CDPSP, *Current Digest of the post-Soviet Press.*

Craumer, P.R. (1994), "Regional patterns of agricultural reform in Russia", *Post-Soviet Geography*, **35**(6), 329–51.

Ekonomicheskoye polozheniye regionov Rossiiskoi Federatsii. Gosudarstvennyy komitet Rossiiskoi Federatsii po statistike (1994).

Fillipov, V. (1993), "Pervyi fermer rossii", *Izvestiya*, 2 October.

IFC (1993), "Privatization of state and collective farms. Nizhny Novgorod. Russia", unpublished manuscript, International Finance Corporation.

Konovalov, V. (1993), "Ot agrarnogo lobbirovaniya stradayet sel'skogo khozayaistvo", *Izvestiya*, 7 August.

Lerman, Zv *et al.* (1994), "Self-sustainability of subsidiary household plots", *Post-Soviet Geography*, **35**(9): 526–42.

Nikol'skii, S. (1993), "Chego zhe khotyat krest'yane?", *Vash Vybor*, 3, 32–3.

Nikonov, A. (1995), "Al'ternativa reforme net", *Khozyain*, pp. 5–6.

Pallot, J. (1990), "Rural depopulation and the restoration of the Russian village under Gorbachev", *Soviet Studies*, **42**(4): 655–74.

Pallot, J. (1991), "The countryside under Gorbachev", in Bradshaw, M.J. (ed.), *The Soviet Union: a New Regional Geography?*, London: Belhaven, pp. 83–100.

Pallot, J. (1993), "Update on Russian Federation land reform", *Post-Soviet Geography*, **34**(3): 211–17.

Petukhova, N. (1993), "Fermery i fermery", *Vash Vybor*, 3, 34.

Prosterman, R.L., Hansted, T. and Rolfes, L.J. Jr (1993), "Agrarian reform in Russia. Report on a policy study and fieldwork in collaboration with the Agrarian Institute, Moscow". *RDI Monographs on foreign aid and development 11*. Moscow.

Rybakovsky, L.L. and Tarasova, N.V. (1989), "Sovremennyye problemy migratsii naseleniya SSSR", *Istoriya SSSR*, 2.

Sazonov, S. (1994), "Tendentsii razvitya fermskogo dvizheniya", *Khozyain*, 7.

Suslikov, S. (1993), "Omskie fermery sozdayut kolkhozi", *Izvestiya*, 6 July.

SWB, *Summary of World Broadcasts* (Daily summary of Russian language broadcasts).

Timofeev, L. (1993), "Apparat protiv kapital. Iskhod etoi bitvy opredelit budushcheye derevni", *Izvestiya*, 16 June.

Unwin, T. (1994a), "Structural change in Estonian agriculture: from command economy to privatisation", *Geography*, **79**(3): 246–61.

Unwin, T. (1994b), "Agrarian change and integrated rural development in a policy vacuum: the case of Estonia", *European Urban and Regional Studies*, **1**(2): 180–5.

Uzun, B. (1993), "Poslednyi mesyats", *Izvestiya*, 3 January.

Van Atta, D. (1993) "Yelt'sin decree finally ends 'second serfdom' in Russia", *Radio Free Europe/Radio Liberty Research Report*, **2**(46): 33–9.

Wegren, S. (1993), "Trends in Russian agrarian reform", *Radio Free Europe/Radio Liberty Research Report*, **2**(13): 46–57.

Wegren, S. (1994a), "Yeltsin's decree on land relations: implications for agrarian reform", *Post-Soviet Geography*, **35**(3): 166–83.

Wegren, S. (1994b), "New perspectives on spatial patterns of agrarian reform: a comparison of two Russian oblasts", *Post-Soviet Geography*, **35**(8): 455–81.

Wegren, S. (1995), "Regional development of Russian peasant farms: A comment", *Post-Soviet Geography*, **36**(3): 176–84.

Williams, G. (1994), "Agrarian reform in Russian and the Nizhny Novgorod land privatisation project, 1990–1994", unpublished M.Phil thesis, University of Oxford.

Yakovleva, E. (1993a), "Kak spasti sel'skoye khozyaistvo, ne uvelichivaya emu dotatsii", *Izvestiya*, 26 March.

Yakovleva, E. (1993b), "Kak idyet agrarnaya reforma v Rossii", *Izvestiya*, 7 October.

9

The post-Soviet environment

Philip R. Pryde
San Diego State University, USA

Introduction

In the period since the dissolution of the USSR in 1991, a high level of environmental deterioration has been well documented to exist in its former republics (Pryde, 1991; Feshbach and Friendly, 1992; Massey Stewart, 1992; Peterson 1993; Wolfson, 1994). Renewable resources such as forests and rivers have been poorly managed and in many areas are seriously degraded. Worse, air, water and land resources over much of the former Soviet Union are all highly polluted, and public health has suffered as a result.

Causes of the problem

The primary cause of this dilemma is generally attributed to the primacy given industrial output during the Soviet period, combined with a low priority for the kinds of environmental protection measures that should have accompanied this economic expansion. It is now clear that maximising production levels is a poor model for a national economic plan, and that a healthy economy cannot be built on top of an unhealthy natural environment. The philosophical belief of Marxist ideologists that there could be no unwise use of natural resources under socialism only added to the problem. Ecologists and natural resource managers understand the adage that "everything has to go somewhere", with the concept of "everything" including effluents from factories, toxic wastes, etc. Unfortunately, Soviet planners always seemed to feel that "somewhere" could be rivers, lakes, oceans, the atmosphere, or anywhere else which in the short run was inexpensive and convenient. Today, the foolishness of treating the environment in this way has become clear.

A related philosophical problem that caused great harm was the view of Stalinist planners that saw nature as a harmful force that needed to be subdued and transformed. As a result, many large-scale projects to "improve" marshes, deserts and rivers were undertaken as a way to implement this questionable goal. Many of these misguided efforts eventually resulted in environmental disasters.

The foregoing considerations, taken together, strongly suggest that there were fundamental problems in the Soviet system of economic and environmental management (Ziegler, 1987). Although Marx and Engels wrote about the necessity of protecting nature, those who described themselves as implementing Marxism were unable to achieve this goal. Part of the reason seems to have been certain poor assumptions, such as assigning no *in situ* value to natural resources, and an erroneous belief that a centrally planned economy would as a matter of course treat nature more gently than an economy based on private enterprise. Another defect in the Soviet system was that those who suffered the most from environmental deterioration, the general public and especially the working population, had no political power to bring about effective environmental changes. Yet it is quite possible that citizen activism may be the most important of all checks on both private and governmental abuses of the environment.

Partially as a result of the above, the Soviet system proved itself over the years to be extremely slow to make substantive reforms in any area. The unfortunate result of all the above was a substantial wastage of the USSR's once-bountiful natural resource base, as well as widespread environmental degradation.

Government policies and inertia notwithstanding, by the start of the 1970s even Soviet sources were acknowledging serious air pollution in industrial cities, water pollution in numerous rivers and lakes, overcutting of commercial forests, widespread poaching of game animals, considerable overuse of pesticides and a variety of other environmental abuses. The extent of these problems, though, was for the most part understated, poorly documented and little publicised. Nevertheless, by the late 1960s, some attention was being given to the most significant ecological threats. At that time, the one issue that highlighted environmental neglect as an issue of national concern was the threatened pollution of Lake Baykal. Although the fate of Lake Baykal is

Geography and Transition in the Post-Soviet Republics. Edited by M.J. Bradshaw.

uncertain even today, it did act as a catalyst to awaken interest in a variety of other environmental problems.

Two events took place in the Soviet Union in the late 1980s which served to focus attention on environmental problems. The first was the tragic explosion at the Chernobyl nuclear power plant, and the second was the initiation of Gorbachev's policy of *glasnost*. The latter permitted both the public and the media to give much greater attention to governmental mismanagement and neglect in areas such as environmental pollution, and resulted in the publication of a number of official reports which discussed the environmental situation inside the country in more detail than had ever occurred previously (e.g. *Natsional'nyy doklad . . .*, 1991).

Some of these reports utilised a new method of analysing Soviet environmental problems, and their relative significance, that being the compilation of maps of critical ecological zones (Kotlyakov *et al.*, 1991). In addition to displaying such zones cartographically, these maps also classified the severity of environmental problems in a given region as falling into one of five descriptive categories: provisionally favourable, satisfactory, stressed, critical (or crisis) and catastrophic. In some cases, the cause of a region's unfavourable classification might be one major problem, such as fall-out from Chernobyl, whereas in other cases the classification might reflect a combination of several types of environmental degradation.

The primary environmental problem regions

Sixteen critical environmental areas were initially mapped, although today a minimum of 20 can be identified, based on information that has come to light in recent years. These critical regions are about equally divided between the Russian Federation and the other 14 former republics of the USSR, and are identified numerically on Map 9.1. The numbers on Map 9.1 correspond to the numbers shown in parentheses in the following paragraphs.

Within the Russian Federation, the primary critical areas include the Moscow region (14), the Volga river (9), the Kola Peninsula (10), the St Petersburg region (17), the Ural Mountains industrial region, including the area of the Kyshtym disaster (11), the Kuzbass coal-mining region (12), the Kalmyk republic (8), Bashkortistan (19) and Lake Baykal (13). Three large degraded water bodies – the Black and Azov seas (6), and the Caspian Sea (7) – are also partially within the Russian Federation. Recent reports about Novaya Zemlya (18), involving the deliberate dumping of nuclear wastes in the adjacent Kara and Barents seas, plus a long history of nuclear weapons testing, suggest that this region should also be designated a critical environmental region.

In the minority republics, the main critical areas are the Aral Sea (1), the Chernobyl fall-out region (2), the Sea of Azov and Black Sea (6), the Donbass region (3), Moldova (5), the Fergana Valley (15), the Dnepr industrial region (4), north-eastern Estonia (17), the Semipalatinsk (Semey) region of Kazakhstan (16), Armenia and western Azerbaijan (20), and the non-Russian portions of the Caspian Sea (7). The Donbass and the Azov Sea problems are shared by Russia and Ukraine, Chernobyl's pollution is shared by these two republics and Belarus, the Black Sea problem is aggravated by Russia, Ukraine and Georgia, the Caspian Sea borders four republics, the Fergana Valley problem is shared by three Central Asian republics, and all five of the latter contribute to the Aral Sea problem.

At the turn of the 1990s, two of the above areas were generally considered by Soviet researchers to fall into the catastrophic category: the Chernobyl region and the Aral Sea region. Information made available subsequent to the collapse of the USSR suggests that the area of the Kyshtym disaster, and perhaps Novaya Zemlya as well, should also be placed into this highest contamination category.

A look at the main problem areas in the Russian Federation

Perhaps the most critical environmental zone within the Russian Federation is the Urals industrial region (11), a large polygon of pollution roughly bounded by the cities of Perm, Nizhniy

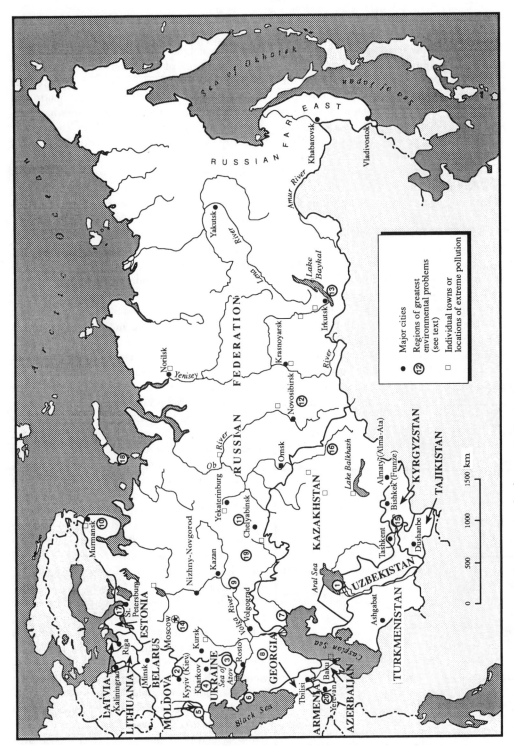

Map 9.1 *Environmental problem areas in the former Soviet Union*

Tagil, Chelyabinsk, Magnitogorsk and Orsk. Here the most widespread problem is vast amounts of air pollution put forth from steel mills, non-ferrous smelters, chemical plants, oil refineries and other industrial operations; but water pollution is also severe. Magnitogorsk has long been used as one of the worst examples of a polluted steel mill town (see Map 9.2). Adding significantly to the regional problem is another area of deadly radioactive contamination east of the city of Kyshtym (the former "secret city" of Chelyabinsk-40), which was the location of at least three major radiation-release accidents in the 1950s and 1960s (Bradley, 1992). These accidents collectively sent far more radioactive waste products into the surrounding land, water and air than did the Chernobyl explosion.

The minority republic of Bashkortistan (formerly the Bashkir Autonomous Republic), which lies in the western foothills of the Urals, is another highly polluted region (19). It is adjacent to the previous region, and is in the lower left portion of Map 9.2. Within Bashkortistan, the main sources of harmful emissions are petroleum

refineries and numerous petrochemical plants, at such cities as Ufa, Salavat and Sterlitamak.

Another area of pollution caused mainly by steel mills is the Kuzbass, an acronym for the Kuznets Basin (12). This valley is a major coal basin, where dusty underground mines have created a significant health hazard for the miners. The large steel mill city of Novokuznetsk has some of the worst air pollution in the former Soviet Union.

In the European portion of the Russian Federation are a number of critical environmental regions. Not exempt from this list is the capital city of Moscow itself (14). Although intended to be a model national capital, the south-eastern portion of the city contains a diverse and polluting industrial region. A more recent scourge is automotive traffic, especially heavy trucks; Moscow and St Petersburg are the only major cities where motor vehicles are a larger source of air pollution than is industry. West of Moscow, areas near the cities of Bryansk and Orel have received unacceptable levels of fall-out from the explosion at the nearby Chernobyl power plant.

Russia's second largest city, St Petersburg (formerly Leningrad) likewise has major problems, though here they are involved more with water pollution (17). The city's water supply, originating in Lake Ladoga and the Neva river, has long been contaminated and is considered nonpotable. To the west, an ill-conceived dike in the Gulf of Finland, intended in theory to protect the city from flooding, has created significant pollution problems in the eastern end of the gulf, and the dike project will have to be redesigned or abandoned.

Another well-known water body facing severe pollution is the Volga river (9). This storied river, often used as a metaphor for the nation of Russia itself, has huge amounts of industrial and urban wastes directed into it. These wastes, in combination with the numerous dams and reservoirs that have been built on it which stagnate water movement, have created many algae-covered sections that are in danger of becoming dead zones. The dams have also eliminated the river's once productive sturgeon runs.

In addition, these same dams have been one of the causes of serious environmental problems for

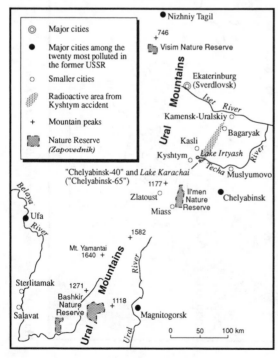

Map 9.2 *The Ural mountains critical environmental areas*

another major water body, the Caspian Sea (7). The effect of the dams was to reduce the flow of the Volga River, which contributes 80% of the total inflow into the Caspian Sea. As a result, for several decades the surface level of the sea declined, causing widespread problems for coastal cities, fisheries and transportation. In recent years, wetter climatic conditions have caused the sea's level to rise a bit.

To the north, the Kola Peninsula is now known to be highly contaminated (10). The main offenders are two very polluting nickel smelters, located at the towns of Monchegorsk and Nikel. In addition, nuclear reactors and nuclear waste storage sites near the city of Murmansk have created a threat of radioactive pollution.

Far greater radioactive pollution exists in the area of Novaya Zemlya (18). These Arctic Ocean islands were initially contaminated by being one of the sites for the USSR's atmospheric nuclear weapons tests, but more recently they (and the waters adjacent to them) have been used as a dumping ground for radioactive wastes of all types. This is causing great concern for marine life all along Russia's Arctic coast.

The Sea of Azov (6) is largely surrounded by the Russian Federation, and receives most of its water from the polluted Don river. It is very shallow, which in the past made it a very productive fishery, but this shallowness has also allowed it to be rapidly polluted. Diversion of Don river water for irrigation has exacerbated its problems.

The final critical area in the Russian Federation is Lake Baykal (13). As noted, it served in the 1960s as a catalyst for the first awakening of the Russian environmental movement. Lake Baykal is the deepest, and perhaps biologically most important freshwater body on earth. The problem derived from excessive timber harvesting along the steep hills that line its shores, and from two polluting pulp mills at Baikalsk and Selenginsk that processes the timber (see Map 9.3). Although some improvements have been made over the past 30 years (logging has been banned, and new national parks and nature reserves have been created), nevertheless the mills still operate, and many scientists are still concerned about the deteriorating water conditions in the lake.

The most critical environmental areas in the minority republics

Within the minority republics, the disaster at the Chernobyl nuclear power station is undoubtedly the best known of the critical environmental problems (2). As a result of the fall-out produced by the 1986 explosion, it has been necessary to relocate over 100 000 persons, billions of dollars have been spent in clean-up costs to date, and it is generally believed that several thousand persons have suffered immediate or premature deaths in Belarus, Ukraine and Russia. Contamination of soils and vegetation from the radioactive fall-out means that large areas of Belarus and Ukraine will be left uninhabitable and unfarmable for decades. Radiation also rained down on several portions of Western Europe and the United Kingdom following the explosion.

Another region of widespread radioactive contamination in a minority republic is the former nuclear weapons-testing site near the cite of Semey (formerly Semipalatinsk) in eastern Kazakhstan (16). Atmospheric testing was conducted there for about 15 years. A short way to the east of the test site is a region of high air pollution levels and heavy-metal contamination, resulting from mining and smelter operations in the Semey (Semipalatinsk) and Oskemen (Ust-Kamenogorsk) mining districts.

The Aral Sea disaster (1) affects a huge area and is almost as well known as Chernobyl. It involves the drying up of the Aral Sea, a situation resulting from excessive withdrawals of water for irrigation from its two main tributaries, the Amu-Darya and Syr-Darya. This diversion of river water has resulted in an Aral Sea greatly reduced in size, the creation of salt-dust storms that originate from high winds that periodically blow across its newly exposed bottom sediments, much damage to nearby cropland, and biological harm to local rivers, riparian habitat areas, and life within the sea itself. Within Central Asia generally, but especially within Uzbekistan's Karakalpak republic, overuse of pesticides and fertilisers and associated contamination of water supplies have contributed to severe public health problems.

The Donbass (Donets Basin) is a coal-mining and steel-producing region located in eastern

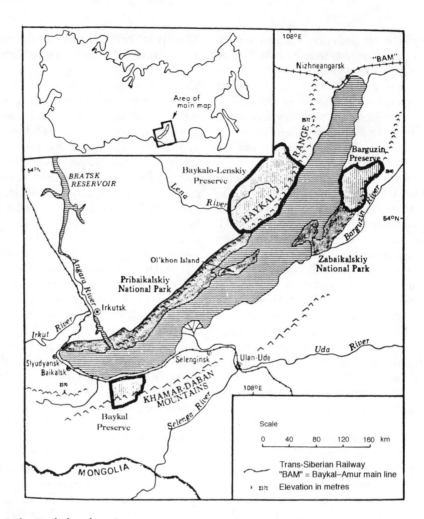

Map 9.3 *Lake Baykel and environs*

Ukraine (3). These activities cause inhabitants there to suffer from extreme air pollution, and only slightly less water pollution. The mining activities also produce a blighted and degraded landscape, and inflict severe health problems on the miners. A short distance to the west lies the Dnepr industrial area (4), also dominated by steel mills and other heavy industry. Its major cities of Kryvyy Rih (Krivoy Rog), Dnipropetrovsk and Zaporizhzhya (Zaporozhye) likewise face the problem of excessive industrial emissions.

In the Kalmyk republic (8), poor agricultural practices have resulted in erosion, loss of soil fertility, and chemical contamination of land and water resources. The same agricultural problems also characterise Moldova (5), with industrial

pollution being a further aggravation that is encountered in several of its larger cities.

The Fergana Valley region (15) is shared by Uzbekistan, Kyrgyzstan and Tajikistan. It is an area of fertile soils that were brought under irrigated farming in the 1930s, and which today have a highly unhealthy mix of not only agricultural chemicals, but also industrial contaminants from mines and processing plants in the surrounding mountains. Some of the mineral resources mined or processed nearby are highly toxic, such as uranium and mercury.

Armenia and Azerbaijan have been engaged in armed hostilities for several years, which has resulted in much environmental destruction in those areas of Azerbaijan near the Nagorno-Karabakh region where most of the fighting has taken place (20). In addition, associated blockades of fuel headed for Armenia have caused forests to be felled for firewood in that republic, resulting in deforestation over extensive areas. Azerbaijan has also been a significant source of pollution (mainly from oil drilling operations) into the Caspian Sea, thereby exacerbating the other problems of that water body that were outlined above.

In a similar manner, industry and agriculture in Ukraine are a contributing factor to the problems of the Black and Azov seas (6), discussed in the section above on the Russian Federation.

The final critical environment to be noted here is found in north-eastern Estonia (17). Here, the mining of phosphorite and oil shale, and associated industrial operations, have produced some of the highest air pollution concentrations to be found in any former minority republic.

Other significant environmental problems

The preceding paragraphs have identified the primary regions of severe environmental disruption within the former USSR, but they by no means cover all the areas of concern. Some environmental problems, such as soil erosion or over-harvesting of timber, degrade large areas that for the most part lie outside of the regions described above. Conversely, other specific nodes of intense pollution occur at numerous discrete points within the former Soviet Union which are also outside the regions outlined above. Among these is possibly the single dirtiest industrial complex on the planet, the giant nickel smelter outside Norilsk in the Siberian Arctic. This one smelter complex in the late 1980s was emitting almost twice as many air pollutants as was the most polluting large industrial city, as indicated in Table 9.1.

Among the other significant centres of pollution are the Irkutsk–Cheremkhovo industrial district along the Angara river, the Kansk–Achinsk coal-mining and power plant complex in east Siberia, the extensive oil extraction region in the west Siberian lowland, the steel production complexes at Mariupol, Lipetsk and Cherepovets, the central Kazakhstan mining and steel mill cities of Temirtau and Ekibastuz, as well as a host of other specific mining/smelting/refining centres.

Table 9.1 *Soviet cities with highest emission levels (emissions in thousand metric tons)*

City	Stationary sources	Mobile sources	Total emissions
Norilsk	2400.1	25.9	2426.0
Krivoy Rog	1290.0	79.2	1369.2
Moscow	369.1	841.5	1210.6
Temirtau	998.9	19.0	1017.9
Ekibastuz	971.4	21.2	992.6
Novokuznetsk	892.9	55.8	948.7
Magnitogorsk	871.4	28.5	899.9
Mariupol	785.8	38.4	824.2
Baku	489.6	297.8	787.4
Lipetsk	722.1	61.6	783.7
Nizhniy Tagil	685.2	26.7	711.9
Cherepovets	671.7	n.g.	671.7+
Leningrad	254.1	371.9	626.0
Omsk	479.4	143.4	622.8
Chelyabinsk	446.7	86.5	533.2
Angarsk	466.8	15.2	482.0
Ufa	349.1	126.4	475.5
Dnepropetrovsk	321.2	122.6	443.8
Zaporozhye	286.9	120.3	407.2
Krasnoyarsk	291.0	108.5	399.5

Source: *Okhrana okruzhayushchey sredy ...* (1989, pp. 22–7).

Air pollution, in general, is a most critical problem in all parts of the former Soviet Union. Heavy industry was developed for half a century with almost no regard to curtailing the huge volumes of harmful emissions that these industries inevitably produce. As a result, almost all large industrial cities have unhealthy atmospheric conditions, some extremely so (refer to Table 9.1). Notice that only in Moscow and St Petersburg (Leningrad) do mobile (transportation) sources exceed stationary (industrial) sources.

In addition to air, water and hazardous materials pollution, a number of other types of environmental deterioration afflict the territories of the former Soviet republics. Depletion of biotic resources, including both flora and fauna, is central among these. In addition to pollution of the landscape, over-hunting and poaching have taken a toll on wildlife in all of these republics. Seventy species of wildlife, including 23 types of mammals and 21 species of birds, have been previously listed as endangered, but this list is widely considered to be incomplete. In an effort to help counter this trend, the various republics have created almost 200 nature reserves, some of them quite large (Braden, 1987). But with only a

few exceptions, the percentage of these republics that are in some status of preservation is fairly small, as can be seen in Table 9.2. Furthermore, these reserves are not always located where the flora and fauna problems are the most acute, and many have difficulty carrying out their tasks due to underfunding and understaffing, compounded in some cases by political neglect and corruption.

Important biotic resources exist elsewhere than just in nature reserves, however. The rich soils of these republics, often high in organic matter, can also be considered to be important biotic resources. Yet wind and water erosion are seriously deteriorating them in many areas, and this results in lowered agricultural productivity and declining crop harvests. Erosion is particularly serious in parts of the fertile steppe and forest–steppe belts of Ukraine, Russia and Kazakhstan.

Likewise, forest resources are also being squandered. Reference was made earlier to over-cutting around Lake Baykal, but this problem is common elsewhere as well, especially in the more populous European portion of the country. A failure to properly reseed cut-over areas so as to assure new forests for the future is a common

Table 9.2   *Preserved land in the former Soviet republics*

Republic	Square kilometres preserved[a]	% of republic preserved
Armenia	3 070	10.30
Estonia	3 311	7.34
Tajikistan	8 862	6.19
Lithuania	3 391	5.20
Azerbaijan	4 407	5.09
Balarus	8 436	4.06
Russia	656 557	3.85
Latvia	2 274	3.57
Turkmenistan	17 184	3.52
Georgia	1 987	2.85
Kyrgyzstan	5 600	2.82
Ukraine	8 111	1.34
Uzbekistan	3 264	0.73
Kazakhstan	12 648	0.47
Moldova	149	0.44
Former USSR	739 251	3.30

[a] "Preserved" lands include formal nature reserves, national parks, and biotic and hunting preserves.

Source: Pryde (1991, pp. 153–8 and 210) (data as of mid-to-late 1980s).

form of neglect. These avoidable losses of resources create not only contemporary problems, but also rob future generations of the resource base they will need for their own economic betterment.

Another critical issue in some of these new nations involves demographic concerns, and one of the most troublesome of these is fertility rates (Joint Economic Committee . . ., 1993, pp. 795ff). Currently, birth rates tend to be highest in the Islamic republics, and lowest in the Slavic and Baltic regions. In the Central Asian republics and Azerbaijan, the combination of high birth rates and insufficient employment opportunities could create unrest among younger people. Efforts to address the underemployment problem could cause the political leadership to subordinate environmental stewardship to a single-minded emphasis on creating jobs. Were this to happen, it would be particularly unfortunate since these Islamic republics represent areas within the former Soviet Union that had some of the highest indices of air and water pollution, pesticide contamination and resultant public health problems.

Economic implications of the dissolution of the USSR for the environment

The gaining of independence by the former Soviet republics will, in general, have unfavourable consequences for the state of the environment in these new nations. Most of the reasons for this involve economic factors that inhibit investment in environmental improvements, such as unemployment, inadequate investment capital, inflation and an impulse to do whatever is necessary to create new economic opportunities. When national economies are experiencing hardships, it is not uncommon for environmental issues to be thought of as an unaffordable dream, and to be deferred to some indeterminant time in the future. But the reality remains that each of these new countries is now responsible for funding its own environmental improvements, in addition to all of its other social needs as well. Unfortunately, in many of them, economic development may well be given a priority over ecological and even health concerns, and environmental in-

dices may at best fail to improve, and in some of the poorer republics might actually deteriorate further from the conditions that prevailed at the end of the Soviet era. To avert this, foreign economic help may be sought in order to achieve environmental improvements, but foreign corporations might be even less concerned about the local environment than would domestic enterprises.

Another economic consideration for each of the new nations will be the availability of key natural resources. Table 9.3 shows that these resources were not evenly distributed across the former Soviet Union. Some republics may find that strategic natural resources that exist mainly in other republics will now cost them much more to import. Rather than accepting the necessity of importing fuels, key metals or timber from Russia or other republics at greater cost than was required under the Soviet economic system, they might endeavour to exploit more marginal stocks of similar resources within their own borders. This might carry a high environmental price.

Other unpopular compromises might have to be made by some republics in the name of economic self-sufficiency. For example, Armenia, Ukraine, Kazakhstan and Lithuania might have to continue operating their worrisome Soviet-built nuclear generating stations in order to avoid energy shortages.

This raises a larger issue relating to nuclear energy: does sufficient expertise exist in these minority nations to satisfactorily carry out their own nuclear power programmes? Lithuania and Armenia have wanted to create a nuclear-free energy base, and Ukraine desires to shut down the Chernobyl complex, but economic realities may require these nuclear facilities to remain in service for some time yet, causing continued nuclear safety concerns. Additionally, will the radioactive wastes of these plants be stored in the republic containing the nuclear facility, or will Russia be willing to accept them for processing at their facilities? And who will pay to dismantle obsolete reactors and clean up existing contamination? Russia, as the successor state to the USSR, is providing minimal assistance to Belarus and Ukraine to help correct problems resulting from the Chernobyl explosion. In addition to nuclear reactors, there exists a wide variety of

Table 9.3 *Economic resources of the former Soviet republics*

Republic	Soil for agriculture	Fossil fuel resources[a]	Commercial nuclear reactor units	Non-fuel mineral resources	Electrical energy production 1990[b]
Armenia	Fair[c]	Few	2	Some	3.17[b]
Azerbaijan	Good[c]	Moderate O, some NG	0	Some	3.30
Belarus	Good	Some NG	0	Few	3.87
Estonia	Fair	Sizeable OS	0	Some	10.93
Georgia	Good	Few	0	Moderate	2.61
Kazakhastan	Very good	Sizeable C, O, some NG	1	Abundant	5.28
Kyrgyzstan	Good[c]	Some C	0	Some	3.12
Latvia	Good	Few	0	Few	2.46
Lithuania	Good	Few	2	Few	7.70
Moldova	Very good	Few	0	Few	3.62
Russia	Fair–very good	Abundant C, O, NG	28	Abundant	7.34
Tajikistan	Good[c]	Some NG	0	Some	3.54
Turkmenistan	Good[c]	Abundant NG, some O	0	Some	4.13
Ukraine	Very good	Sizeable C, NG, some O	15	Sizeable	5.77
Uzbekistan	Good[c]	Sizeable NG, some C	0	Some	2.83

[a] C = coal; O = oil; NG = natural gas; OS = oil shale.
[b] In 1000 kWh per capita; in 1990 the average for the USSR as a whole was 6020 kWh/per capita.
[c] If irrigated.
Sources: *Atlas SSSR*, 1983; Shabad (1969); Jensen *et al.* (1983); *Post-Soviet Geography*, **32**(4) (April 1991), and **33**(4) (April 1992).

other sources of radioactive contamination throughout the former Soviet Union, as summarised for each of the republics in Table 9.4.

Similar concerns exist for several other industries that are inherently "dirty", such as iron and steel, petroleum refining, chemicals, and fossil fuel power plants. Table 9.1 illustrated the problem that many cities have because of such enterprises. Many republics have high aggregate emission statistics (Mnatsakanian, 1992). Each of the new nations will need to effectively control air and water emissions and toxic wastes from all their dirtier industries, and this will have significant budgetary implications for these countries.

Nor is a "new generation" of more far-sighted industrial leadership apt to appear any time soon; for the most part the same middle-management personnel that made up the ineffective regulatory process under the old Soviet ministerial system are still in office.

Another cause of future environmental problems might be inter-republic competition. This could result if these new nations vigorously compete for foreign economic investments and new employment centres, using relaxed environmental regulations as part of the enticement package, in order to make their republic appear more "business-friendly" than their neighbours. This

Table 9.4 *Sources of radioactive wastes*

Republic	Nuclear power plants	ICBM missile sites	Surface bomb or accident residue*	Under-ground atomic tests	Mining, milling or processing sites	Nuclear research centre	Radioactive waste burial sites
Armenia	X					X	X
Azerbaijan							X
Belarus		X	X			X	X
Estonia					X		X
Georgia						X	X
Kazakhstan	X	X	X	X	X	X	X
Kyrgyzstan					X		X
Latvia						X	
Lithuania	X						X
Moldova							X
Russia	X	X	X	X	X	X	X
Tajikistan					X		X
Turkmenistan				X	X		X
Ukraine	X	X	X	X	X	X	X
Uzbekistan				X	X	X	X

* Residual contamination from either atmospheric nuclear tests or major accident.
Sources: Derived from material in Bradley (1992) and Potter (1993).

is a common game among states and cities in the United States, and could likewise become an expedient and counter-productive tactic among the former Soviet republics. A related question involves whether each of these new nations will have the internal organisation (and political will) to supervise and regulate adequately the development of their natural resources by foreign companies. An early case study involved the leasing of timber resources in Russia's Maritime Province, and there the answer to this question seemed to be "no" (although pressure was later put on the regional officials to give more weight to environmental protection).

The nationalistic concerns of minority peoples and regions also have environmental ramifications, a problem that can most easily be seen in the giant Russian Federation. Here, the largest individual subunit, the former Sakha Autonomous Republic (now known as the Republic), has demanded greater local control over its huge storehouse of natural resources. In a similar manner, Tatarstan (the former Tatar republic) has indicated that it would like a greater amount of political sovereignty as well. If most of the 20 minority republics within Russia were to act in a similar way, this might result in much of the natural resource riches of the Russian Federation being controlled by local autonomous republic governments, rather than by Russian government agencies in Moscow. Their decisions concerning the development of these resources might not be to the liking of the Moscow leadership, and might not necessarily be environmentally sound.

International relations and the environment

A universal environmental problem that has been made more complex by the creation of many new smaller countries is the dilemma of transboundary pollutants. There are several components to this problem:

1. pollutants that move through international lakes or river systems;
2. airborne pollutants, which include in addition to smokestack and exhaust emissions such other contaminants as windblown pesticides, salts and other harmful solid matter;
3. pollution of international seas (such as the Black, Baltic and others) whose coastlines take in portions of two or more countries;

4. internationally conveyed pollutants, meaning those that are legally moved (or in some cases illegally smuggled) across international borders.

Many of these new countries are relatively small in size, and therefore transboundary pollution can be an issue that affects a number of them. They also may not have the financial resources to attempt to alleviate these transboundary incursions. An assortment of new international accords may be required to address these problems.

In the early 1990s, several armed conflicts were taking place, or had recently occurred, in portions of the former Soviet Union; and such hostilities almost always produce considerable environmental destruction. The most serious conflict, which started in the late 1980s, has been in the Nagorno-Karabakh autonomous region and surrounding portions of Azerbaijan. Other areas which have had armed disputes include the Ossetian and Abkhazian minority regions of Georgia, as well as sporadic civil–political conflicts in Tajikistan. Not only is there direct environment damage brought about by such hostilities, but the high monetary outlays for the military activities often precludes funds being available for environmental restoration even after peace has been achieved. In many cases, decades many have to pass before the environment returns even partially to pre-war conditions.

Even if they are not being actively used, modern military weaponry, such as chemical and nuclear weapons, represent a serious long-term problem. In the case of atomic weapons, the cold war had hardly ended when, at the start of 1992, the number of nations possessing strategic nuclear weapons and long-range delivery systems suddenly increased by three. As a result, the scope of the ongoing strategic arms reduction negotiations had to be enlarged to include Belarus, Ukraine and Kazakhstan, in addition to Russia. The problem is not so much that any of these countries would actually use the weapons, as it is that the weapons are deteriorating and the smaller countries have no facilities to handle the radioactive materials. A related question involves how the United States and Russia will dispose of the conventional explosives, enriched uranium, rocket propellants, ship-borne reactors, chemical

weapons and other highly hazardous war materials that will still exist even after military threats have ended. Disposing of such materials safely will constitute one of the major environmental (and ethical) challenges of the post-cold war era.

The 15 former republics must now each find their economic and environmental niche within the world community of nations (Cole, 1991). An important part of this will be the establishment of new international environmental agreements among themselves and with other countries. As noted above, the potential exists for this to devolve into a lack of co-operation among these, fuelled by competition and misguided self-interest. Hopefully, though, they will see the necessity of environmental co-operation. An excellent example of the type of new multilateral environment agreements that will have to be developed involves the water problems of the Aral Sea basin. Prior to 1991, any decision involving water problems of the Aral Basin would have been made in Moscow, and accepted by all subordinate regional authorities. Now, five independent nations are involved (six, if Afghanistan is properly included in the Aral Basin composition), and a comprehensive basin compact will be needed in order to effectively manage these water supplies in the years ahead. The first step in this direction occurred in 1992 when the five Central Asian republics signed an agreement regarding the management and protection of the basin's water supplies, creating a joint commission (composed of the water resources minister in each republic) to regulate and conserve the water supplies in inter-republic rivers and lakes. This was only an agreement to co-operate, however; specific action plans must still be awaited.

Alternatively, environmental co-ordination would become a central concern of the Commonwealth of Independent States (CIS), which is composed of all of the former Soviet republics except those in the Baltic region. Thus far in its young history, however, the CIS has not demonstrated itself to be a strong policy-implementing body. Under its direction, an agreement entitled "Interaction in the Field of Ecology and Environmental Protection" was agreed to by its member states. This document, like the one on the Aral Sea mentioned above, resembles more a goals statement than a mandate to carry out specific environmental actions. If the CIS remains a weak implementing body, the needed environmental actions could also be accomplished by bilateral and multilateral agreements taken outside the auspices of the CIS.

Summary

The new nations that have evolved out of the former Soviet Union are facing a huge array of developmental challenges. Most have serious economic problems. In this context, it will be tempting for them to put economic considerations ahead of environmental ones, but in the long run this could prove to be a very unwise choice. The environment in many of the former Soviet republics is presently so contaminated that it is adversely affecting other important developmental priorities, such as agriculture, biodiversity, public health and the availability of clean water for new industry and towns. It is self-defeating for these nations to believe that they can construct a healthy economy at the expense of their natural environment, for the latter in all countries is the basis for the former. Environmental decay merely creates additional economic burdens for the citizenry, whose future generations will have to pay these delayed bills, generally at inflated prices. As an additional insult, in the interim the public may suffer severe declines in levels of public health due to this environmental neglect. This is a process that can be clearly seen at present in the Chernobyl fall-out region and in Central Asia. The former Soviet republics must perfect ways of incorporating contemporary costs of environmental maintenance into the prices of the goods they produce on a regular basis, if they wish to avoid transferring today's serious environmental problems to their own posterity.

In several places in this chapter, the factor of limited financial resources has been stressed. It is true that the health of the economy and the environment are related, and that environmental enhancement requires large capital outlays. Unfortunately, none of the republics is well endowed with capital resources at present. Although all of these new nations are seeking outside financial assistance to jump-start their struggling economies, environmental improvements are not usually among the highest priorities. The most optimistic scenario may be that many of the most polluting industries in these republics will permanently close, and the foreign aid can be used to construct more modern plants that make use of cleaner technologies and processes. Even if this occurs, it will take time for significant environmental enhancement to be realised, and in the interim it is most probable that existing factories will continue to pollute the air and waters of these republics. In general, it is unfortunately true that people in almost all of the former Soviet republics prefer smog and poor health to poverty and unemployment (Baiduzhy, 1994).

In all of this though, it should not be forgotten that one other important set of players exists in the former Soviet republics. These are the many private, non-governmental environmental organisations (or "NGOs" as they are often called) that have arisen since the late 1980s. Perhaps the most impressive and best organised among these at present is the Socio-Ecological Union (SEU), headquartered in Moscow. The SEU is the only post-Soviet environmental group that could be called an effective national organisation. Founded before the dissolution of the Soviet Union, it has been able to maintain effective ties with local environmental groups in virtually all parts of the former Soviet Union. Many similar organisations exist at the local and regional levels as well, often focused on one particular problem of major local concern. As a general rule, however, they suffer from insufficient resources and are therefore very often (but by no means always) unable to bring about the desired environmental improvements. One area where these NGOs are active, and where much more work is needed, is in the realm of environmental education. One reason that so little seems to be accomplished environmentally in these republics is that the citizenry is often uninformed as to the reasons why environmental actions are necessary. Usually, ministry officials are even less aware (or concerned). Better environmental education is urgently needed as a catalyst to foster more effective citizen activism and consequent governmental improvements.

The immediate prospects for environmental improvement in the republics of the former

Soviet Union are, unfortunately, not good. The deteriorated environmental conditions that the new nations inherited from the USSR are compounded by the ongoing economic crisis which affords no supply of funds that can be devoted to the needed clean-up. Worse, their administrative organisations often are composed of a managerial élite that still employs many of the same bureaucrats and counter-productive attitudes that caused the problems in the first place. Political reform, public activism and an economic upturn will need to go hand in hand in order to lay the groundwork for the environmental improvements that are so urgently needed in all these new nations.

References

Baiduzhy, A. (1994), "Should we forget about ecology?", *The Current Digest of the Post-Soviet Press*, **46**(11): 1–2.

Braden, K. (1987), "The function of nature reserves in the Soviet Union", in Singleton, F. (ed.), *Environmental Problems in the Soviet Union and Eastern Europe*, Boulder, Colo: Lynne Rienner.

Bradley, D.J. (1992), *Radioactive Waste Management in the Former USSR*, Vol. III. Richland, Wash.: Pacific Northwest Laboratory.

Cole, J. (1991), "Republics of the former USSR in the context of a united Europe and new world order", *Soviet Geography*, **32**: 587–603.

Feshbach, M. and Friendly, A. (1992), *Ecocide in the USSR*. New York: Basic Books.

Joint Economic Committee of the Congress (1993), *The Former Soviet Union in Transition*, 2 vols. Washington, DC: US Government Printing Office.

Kotlyakov, V.M. *et al.* (1991), "An approach to compiling ecological maps of the USSR", *Mapping Sciences and Remote Sensing*, **28**: 3–14.

Massey Stewart, J. (1992), *The Soviet Environment: Problems, Policies, and Politics*. Cambridge: Cambridge University Press.

Mnatsakanian, R.A. (1992), *Environmental Legacy of the Former Soviet Republics*. Edinburgh: Centre for Human Ecology, Univ. of Edinburgh.

Natsional'nyy doklad SSSR k konferentsii OON 1992 goda po okruzhayushchey srede i razvitiyu (1991), Moscow: MinPriroda.

Okhrana okruzhayushchey sredy i ratsional'noye ispol'zovaniye prirodnykh resursov v SSSR: statisticheskiy sbornik (1989), Moscow: Goskomstat.

Peterson, D.J. (1993), *Troubled Lands: the Legacy of Soviet Environmental Destruction*. Boulder, Colo.: Westview Press.

Potter, W.C. (1993), "The future of nuclear power and nuclear safety in the former Soviet Union", *Nuclear News*, March: 61–7.

Pryde, P.R. (1991), *Environmental Management in the Soviet Union*. Cambridge: Cambridge University Press.

Shabad, T. (1969), *Basic Industrial Resources of the USSR*. New York: Columbia Univ. Press.

Wolfson, Z. (1994), *The Geography of Survival: Ecology in the Post-Soviet Era*. London: M.E. Sharpe.

Ziegler, C. (1987), *Environmental Policy in the USSR*. Amherst: U. of Massachusetts Press.

10
Economic restructuring and local change in the Russian Federation

Alison C. Stenning
University of Birmingham, UK

At such a high level of generality, many problems lose their significance, cease being relevant, vanish. The ideological and the national macroscale marginalizes and invalidates the difficult, vexing microscale of everyday life (Kapuscinski, 1994, p. 308).

Why local change?

While the processes of change and transformation in Russia have attracted considerable attention in both the popular and academic media, the predominance of debates at the national and macroeconomic scale has diverted attention away from the spatial aspects of systemic transformation. Many of the processes of transformation are inherently geographical – the reconfiguration of centre–province relations, mass migrations, the (re)construction of new state formations, the (re)defining of identities and citizenship, for example, and many more are inevitably mediated by spatial contingencies – yet, economic transformation has all too frequently been represented as a homogeneous and homogenising process. (Here the debates seem to mirror some Western discussions of globalisation which suggest that increasing internationalisation, and thus an ever wider reach of the forces of marketisation, is heralding "the end of geography" (see O'Brien, 1992).)

Not only do processes of change take place unevenly, but some of them may even have a significant local level of operation – for example, the organisation of labour. Taylor (1982) draws our attention to a political economy of scale in which he sees the global as the scale of reality, defined by the capitalist mode of production which is all-embracing and organises life on every scale below the global. Next comes the national scale, i.e. the ideological and macro scale at which much formal politics and policy making is carried out. Finally, the urban, or local, scale is the scale of experience. Whatever occurs on the real and ideological levels, the effects are lived locally. These are the spaces of everyday life, though what happens at this scale is strongly influenced by processes of change at the global and national levels. This is not to argue that the local is concrete and the rest abstract (see Massey, 1994) but to encourage an understanding of why patterns of transformation and local change (whatever their form) are differentiated and how different scales of change are interconnected. The idea of a scale division of labour is furthered by Cox and Mair who move on from those "theoretical approaches which equate the local with the concrete, the individual, the particular and the contingent, and equate the global . . . to the abstract, the social, the general and the necessary" (1991, pp. 197–8). They argue that any locality must be considered within the wider context of the geographical divisions of labour, both of scale and space, and that each activity primarily associated with a certain scale may indeed be witnessed on other scales. The structure of global–national–local relations is more complex than Taylor suggests but, however the local is constructed, it remains for many the "difficult, vexing microscale of everyday life".

Turning to the former Soviet Union and East Central Europe (FSU/ECE[1]), David Stark argues against essentialist understandings of change and suggests that we are witnessing

a plurality of transitions in a dual sense: across the region, we are seeing a multiplicity of distinctive strategies; within any given country, we find not one transition but many occurring in different domains – political, economic and social – and the temporality of these processes are often asynchronous and their articulation seldom harmonious (Stark, 1992a, pp. 18–19).

Though Stark does not make the point explicitly, the logical conclusion of his plurality of transitions is that change will itself vary through space on every scale, not just between countries. Further, instead of believing that the building of a new order starts from scratch, Stark recognises that "choices are constrained by the existing set of institutional resources"[2] (1992a, p. 21). This is the concept of path dependence which implies that current choices or strategies will to a significant extent be determined by previous paths. This concept of path dependence parallels

Geography and Transition in the Post-Soviet Republics. Edited by M.J. Bradshaw.

somewhat Doreen Massey's "geological metaphor" (see Massey, 1984). Massey argues that we are witnessing national and international processes at work producing a variety of unique local economic and social structures. Historical conditions at the local level mediate the operation of those processes (Massey, 1983). Each layer, or round, of investment and deinvestment, class, gender and race politics, cultural institutions, and so on, on a multitude of scales, are sedimented on to existing structures and institutions to affect the development of capital.

Transition or transformations?

Stark has another criticism of many studies of transformation. Although he himself continues to use the term "transition", Stark argues against the use of phrases such as "the transition to capitalism" or "the transition to a market economy" since they are teleological constructs which assume an end-state and in doing so narrow the field of debate. Any changes are defined in terms of the transition to capitalism and judged according to their utility in getting closer to that endpoint. For example, the economic strategy chosen by regional policy makers in Ul'yanovsk, rather than being seen as a positive alternative which ensures lower prices for consumers, is derided as "Brezhnevian" in comparison to "the strong liberalisation and privatisation approach of St Petersburg and Nizhny Novgorod" (Hanson, 1994b, p. 2). It has been argued that the end (and failure?) of Sovietology has led to the "renaissance of modernization theory" (Burawoy, 1994, p. 774), as claims arise "that progress follows a *single* course toward a market economy and political democracy" and is "failing to come to terms with the specificity of the Soviet experience and the enormous obstacles to development" (Burawoy, 1994, p. 774, emphasis added). There is an assumption that the introduction of market institutions such as private property and a stock market will inevitably lead to a fully fledged market economy. While in 1993 World Bank officials finally admitted that "Eastern Europe is not well served by straight textbook advice" and that "it had been a mistake to assume that state enterprises and banks would behave

according to market principles" (cited in Ellman, 1994, p. 2), assumptions of teleological development still linger in much literature. It may be true to say that FSU/ECE is making some (often quite strong) moves towards some form of capitalism, it is neither *necessary* nor *inevitable* that that form will be as Western advisers and commentators wish.

Without arguing against the pervasive powers of capitalism as a (dis)organising system, Duncan writes that "social (and natural) systems are not universal, they have geographies, histories and contexts. Hence originating, generative mechanisms will not be universally present or equally developed" (1989, p. 139). The modernisation approach exhibited a clear preference for a particular form of capitalism, which was in practice "very much the same as Westernization, i.e. the underdeveloped country should imitate those institutions that were characteristic of Western countries" (Blomstrom and Hettne, 1984, p. 21). More recent, critical literature on social systems argues that they are embedded within the culture and institutions of a nation (region, locality) (see for example Granovetter, 1985). Hutton states that although the profit motive may be universal to capitalism "it takes place within inherited social and political boundaries" (1995, p. 24). Socioeconomic systems work *as a whole* and can only be fully comprehended through an understanding of the "institutional and cultural difference between capitalisms" (Hutton, 1995, pp. 257–8), that despite similarities in the social and economic purpose of capitalist structures and institutions, varying cultures and histories produce significant differences in actual forms.

Capitalist elements have undoubtedly developed in Russia, but it has been argued that these developments rely heavily on the structures of the former administrative-command system (Clarke, 1992). Though relations of exchange and distribution may have changed dramatically (Burawoy and Krotov, 1993), old production relations persist. Politics and economics are still intimately fused and production still depends considerably on political connections and influence. What may be emerging is a form of merchant capital, or political capitalism, "which does not have its own distinctive system of production but grafts itself onto pre-existing systems without necessarily altering them" (Burawoy and Krotov, 1993, p. 54). Chomsky describes such a process as the "Third Worldisation" of the FSU (Chomsky, 1994, p. 152), as IMF orthodoxy and Yeltsin reforms combine with the legacy of state socialism to create "a careful blend of Stalinism and the 'free market'" (Chossudovsky in Chomsky, 1994, p. 152). A small part of the population is enriched as old names are preserved in positions of authority and management, pre-existing distortions are reinforced and foreign capital emerges as a party to this new mercantilism. In this way, a market economy has been able to emerge *without* the restructuring of capital.

Thus, an understanding of the processes of change in FSU/ECE must incorporate an acceptance of pluralities of change, of *transformations*, rather than *a transition*, and must recognise the social, political and cultural context of economic change and capitalist development. The chapter now returns to the focus of a more narrowly defined economic transformation and examines what outcomes are expected. The limited scope of such a focus is considered and it is suggested that through making use of concepts developed to study restructuring in the West a more holistic understanding may be gained.

The four pillars

Starting from the assumption that the former state socialist countries must make the transition to capitalism, as defined by Western advisers and commentators, many writers have recognised four policy "pillars" – liberalisation, stabilisation, privatisation and internationalisation – whose straightforward application should lead to the development of liberal democracy. The first question to answer is why, in theory, these policy processes should lead to economic restructuring.[3]

Liberalisation fundamentally involves the abolition of regulations and restrictions on economic activity. The most crucial example in the post-Soviet context is the freeing of prices. Under the Soviet system prices were hugely distorted and played almost no function at all in controlling supply and demand. Prices did not reflect market value, even within the Soviet

Union. A comparison with world market prices would have been even more devastating. As a result the structure of the economy was vastly distorted. Producer goods were prioritised over consumer goods, with the defence industry playing a peculiarly large role, and large sectors of the economy involved in "purely socialist" production (Hanson, 1994a) which added little to the welfare of the people. In freeing prices, and thereby dramatically reducing the subsidies available to enterprises, the reformers theoretically inflicted harsh market logic on to enterprises and their managers. The need to pay attention to prices and costs demands a restructuring of capacity. Not only may enterprises realise that their products are not competitive in terms of quality, but the extensive mode of industrialisation that was pursued in the Soviet Union becomes unsustainable as wastage – of labour, capital and raw materials – begins to cost.

A macroeconomic disadvantage of the freeing of prices, from previously undervalued levels, is the potential for a sustained process of inflation. There is a distinct possibility that price increases will lead to a price–wage–price inflationary spiral that would undermine the currency. For this reason any programme of rapid liberalisation must be linked to one of *stabilisation*. Budget deficits must be decreased, subsidies cut and credit supply tightened, all at a time when enterprises are facing hugely increased competitive pressures. Without state assistance in such a dire situation the pressure to reform or close should be increased. A survey of 3000 East German companies in 1991 concluded that just 30% were competitive as they stood, 50% needed significant internal restructuring and 20% were "doomed" (Blanchard *et al.*, 1992). It is estimated that about 20% of post-Soviet firms are so uncompetitive on world market terms that they actually destroy value, they are "value subtracting", such that the output of the production process is valued less than the inputs (see McKinnon, 1991, Ch. 12).

The picture is complicated still further by the simultaneous process of *privatisation*. Linked to stabilisation and the reduction of state subsidy and support for enterprise, the explicit aim of a policy of privatisation is not only to shift ownership, and hence responsibility, to the private sector, but also to specifically promote efficiency and restructuring. Privatised firms in a liberalised economy should be acting according to profit motives and aiming to maximise income and minimise costs. The Economist Intelligence Unit notes that by August 1994, 106 000 Russian enterprises of all sizes and sectors, employing 86% of the workforce, had been privatised (EIU, 1995, p. 29). Given the uncompetitive state of much Soviet industry, such a shift to capitalist ownership should, according to neo-liberal theory, have triggered off a very significant process of restructuring.

The final part of the transformation equation is *internationalisation*, a bundle of processes that impinges on all three of the other pillars, in many respects making their effects even more severe. Not only must prices be freed domestically, but they must also now conform to international levels as the Russian economy is opened to the world. Stabilisation must be carried out with the help and hindrance, financial and otherwise, of the international community, particularly the IMF and World Bank, whose harsh structural adjustment programmes are renowned for their tight monetary policy. Russia is not only attempting transition to a market economy, but to a market economy during a frenetic period of speculative capitalism, which threatens to endanger further attempts at stabilisation. Privatised firms must face not only domestic competition, but must also be confronted with competition, from imports and direct investment, from considerably more dynamic, flexible and advanced western producers.

It could have been expected that the policies and exigencies of transformation would have led to a very significant period of internal enterprise restructuring leading to bankruptcies, rationalisation, the introduction of new technologies, the divestiture of numerous sectors and even the relocation of enterprises. However, the evidence points to the contrary. In a study carried out in 1993, enterprises were found to be developing "an orientation towards survival [because] uncertainty of the future did not allow them to pursue a long-term model of development" (Dolgopyatova and Evseyeva, 1995, p. 320). A KPMG study of 27 privatised firms in Russia (cited in Nuti, 1994), claims that

approximately half had made no changes to their operation, that employment and welfare provisions were still prioritised and that few had planned a commitment to restructuring. "Preserving the workforce" was listed as a primary objective by 58% of the enterprise directors interviewed by Dolgopyatova and Evseyeva. Yet, 55.6% recognised the need for a "sound financial position". Furthermore, the "classic market motivation" of "increasing the profitability of production" was acknowledged by a significant 29%.

Despite the fact that as many as one in five enterprises, on most estimates, could be described as insolvent, just eight enterprises were reportedly declared bankrupt in 1993 (Hanson, 1994a). By early 1995, the courts had placed over 500 newly privatised firms in receivership (*The Economist*, 1995, p. 13), and over 4000 more have been declared insolvent (*Russian Economic Trends*, 1995), unable to pay their workers and producing little. The Federal Bankruptcy Agency only has the power to liquidate wholly state-owned enterprises. The procedure for dealing with all other insolvent enterprises is not clarified, though a complex process of judicial arbitration has meant that just 49 state-owned firms have been liquidated while a number of "insolvent" firms have managed to repay debts and function profitably once again. However, the delays and confusion over the bankruptcy procedure suggest that numerous potential closures and redundancies have simply been postponed. Meanwhile, official unemployment is being kept artificially low as apparently solvent enterprises put workers on short-time, or involuntary leave (so-called "hidden unemployment").

There are a number of reasons why enterprise restructuring is not automatically producing free market capitalism. Nuti (1994) sums up the barriers to restructuring by noting that a simple transfer of ownership to private hands is not enough. Much of the privatisation that has taken place in Russia has been defined as "spontaneous", leading to an ambiguity of ownership. Under 1988 and subsequent legislation, 51% of shares could be bought by existing workers and managers on preferential terms.[4] These "new" owners had interests in maintaining domestic monopolies and protection from imports and foreign direct investment, in sustaining subsidies from government and in having the freedom to restructure as and when they wished. Workers fought against job losses and managers, who according to the veto of soviets were accountable to the workers, succumbed to such demands in order to hold on to power. The insider character of much privatisation led to attempts to preserve the status quo. Moreover the lack of a developed market infrastructure, of legal frameworks in particular, has meant a distinct absence of legal certainty of ownership. While workers and managers frequently own a majority of the shares in an enterprise, responsibility for survival still appears to rest with the state. The persistence of what Kornai (1992) calls "soft budget constraints" has meant that enterprises continue to run at a loss, safe in the knowledge that the state will in the last analysis bale them out. Kornai adds that as long as connections can be used to protect enterprises from hard budget constraints and the threat of bankruptcy and liquidation, the behaviour of managers will remain ambivalent.

Legal ownership may change overnight; deeply ingrained attitudes will not. The paternalism that provided for the bail-out of failing enterprises lives on. Two years on from Kornai's analysis "commercial banks and state enterprises are mostly not profit-driven, hard-budget constraint business units . . . there is none of the pressure to innovate that is created by sink-or-swim competition under capitalism" (Hanson, 1994a, p. 17). Government subsidies have continued and, as noted above, bankruptcies have, up till now, been rare.

Restructuring and transformation

That the four processes of transformation described above have not created the liberal democratic haven promised suggests the interplay of forces not conceived of in a narrowly economic view of restructuring. It has been noted that most studies of systemic transformation have focused on economic transition which is in fact "but one component of a more complex process" (see Chapter 1). Yet, it could be argued further that studies of economic transition itself have been severely limited to studies of the economic

aspects of transformation in the economic sphere. Any study of economic transformation in the FSU demands an examination of wider patterns of social change. As Lovering indicates in debates on Western restructuring, "accumulation is by no means confined to the purely economic" (Lovering, 1989, p. 207). Not only is transformation also, and indistinguishably, a political and social process – in essence, the introduction of capitalist social relations – but economic restructuring itself necessarily involves the reconfiguration of political compromises, shifts in work culture, the desocialisation of the workplace, and perhaps most fundamentally of all, processes of individual, enterprise and regional survival. The enterprise, particularly the old-style Soviet enterprise, is not simply an economic structure and its restructuring is not simply an economic process.

John Lovering defined restructuring as a "qualitative change from one state, or pattern of organisation, to another . . . qualitative changes in the relations between constituent parts of a capitalist economy" (1989, p. 198). Restructuring can be seen as the way in which economic actors respond to a changing competitive environment and, consequently, the way in which the organisation of economic activity changes across geographical space. The changing political–economic environment of the 1980s was manifested through a vast range of outcomes, many of which were not simply inherently spatial (e.g. relocation to cut costs) but also varied considerably through space and time. However interesting it may have been to study local outcomes and, in the manner of industrial location theorists, conclude upon location factors (positive and negative), and their relation to industrial geography, this was not enough. Within the restructuring approach (which could by no means be represented as a singular, coherent approach), case studies of economic geography should serve "as windows onto the processes beyond", to aid an understanding of the processes of change. Further, Massey (1988, p. 60) writes that "the restructuring school argued that it was not possible to provide an explanation in terms of locational factors alone", but that it is necessary to set changes within the wider context of the global restructuring of industry. Regional change has to be discussed in the context of capitalist accumulation and the changing demands of its perpetuation. The restructuring school is characterised as an attempt "to elucidate the interaction of capital's strategies and the socio-spatial pattern of production" (Lovering, 1989, p. 218).

In his edited volume *Localities*, Cooke (1989) focused on the local experiences of economic restructuring, social adjustment and political change, specifically the coping mechanisms and survival strategies of enterprises and localities during Britain's deindustrialisation. It was recognised that such changes could create both opportunities and crises, through a Schumpeterian process of "creative destruction". Cooke recorded a number of pressures to change. Perhaps the most influential was increasing international competition, from the newly industrialised countries (NICs) in particular, to a certain extent encouraged by technological change which facilitated international movement and communication. A secondary pressure was linked to the hegemonic position of neo-liberal, monetarist discourses, demanding stable, balanced budgets and low inflation, internal competition in the private sector and encouraging not only the introduction of performance indicators, but also fully fledged privatisation in the public sector. Finally, tendencies to falling rates of profit, crises of accumulation, as productivity growth in manufacturing slowed, shifted the emphasis of economic growth on to new sectors, particularly service industries which provided different types of job, demanded different types of skills and had different geographies.

In response to these pressures, firms could be seen to be adopting a number of restructuring strategies.[5] The most fundamental of all are the traditional policies of rationalisation of labour and capacity, that is redundancies and closures, in the face of the need to cut costs. Firms sought to relocate in semi-rural locations with cheaper labour and rent, and less organised workforces. The technological change that permitted the decentralisation of manufacturing also encouraged the spatial separation of the functions of conception and execution as R&D, and other services, became increasingly located in the metropolitan cores – a new spatial division of labour was emerging. There was an increase in subcontracting and

widened supply networks as enterprises attempted to respond more flexibly to apparently segmenting markets. Moreover, firms began to co-operate as well as compete with their rivals to consolidate opportunities for research and marketing, as these and other auxiliary services took on more and more importance in economic structures. New relationships were also forged between the private sector and government, both central and local, as public–private partnership became the watchword of local governance. The emergence of these new partnerships has been linked to the rise of so-called "place marketing" as conflict within localities is superseded by competition between them, and heterogeneous coalitions[6] within localities offer capital all sorts of incentives to guarantee "their" places a role in the future space-economy.

In short, in response to increasing international integration, to pressures to liberalise, stabilise and privatise the economy and to technological change, firms, and hence entire economies, adjusted and restructured. This was characterised by many as a transition from Fordism to post-Fordism (see for example Amin, 1994). It was certainly a period of profound change, a "shifting kaleidoscope" of change as the interaction of "geographical surfaces" and "the demands of industry" (Massey, 1994, p. 110) were played out in new ways. Each locality is locked into the changing international political economy in very different ways, mediated through a variety of national and continental contexts, such that different kinds of economic, social and political change are going on in different localities.

Along with this belief in contingency of outcomes, work within the restructuring school is inherently committed to the idea that economic restructuring is necessarily bound up with social and political change. Unlike the narrow economic use of the term "restructuring", within geography this phrase refers to much more than the reprofiling and rationalisation of individual enterprises. While perhaps still focusing on enterprises as the agent of change, this approach makes the necessary link between changing competitive structures, changing demand for labour and technology and shifting structures of production and reproduction. The narrow restructuring of enterprises necessarily entails shifts in employment, in technology, in local embeddedness – all bound up with the survival of the local political economy. It is not simply the enterprises that are restructuring but also the structures, institutions and social relations of entire places, be they localities, regions or states.

The use of theoretical approaches developed in the West for the study of "other" places is, however, contentious. With specific reference to the FSU, it seems to imply that now the republics have finally accepted the "superiority" of capitalism as a (dis)organising system, it is possible to uncritically translate theories of capitalist development which have been used for the study of advanced capitalist states for years. In short, it could be seen that the application of Western theories of restructuring to studies of the FSU shows a lack of sensitivity to historically and geographically specific social systems. The issues and concepts of restructuring were developed for the study of the transition from Fordism in Western late capitalist states. However the states of the FSU are characterised as "state socialist", "state capitalist" etc., it is difficult to describe them as "late capitalist". The debate over the parallels between post-Fordist and post-communist transformations is still relatively undeveloped (see Peck and Tickell, 1994 for the beginnings of debate), but it is clear that there are both similarities and differences in these processes. What is important in Western work on restructuring is the conceptualisation of regional change as the interplay of "big" processes and local conditions, and a recognition that the same mechanism can and will produce different outcomes, and, conversely, that the same effect can and will result from different causes. For example, the need to cut costs could lead to either a relocation or rationalisation of the workforce (or both), while a rationalised workforce could also be the result of the introduction of new technologies. To understand spatial change on every scale, from the international to the local, demands an understanding of the "contextual conditions which can affect how or if causal powers act" (Duncan, 1989, p. 133). Contextual conditions – that is, patterns of historical development, particular moments in institutional relations – exist on every spatial scale to mediate the development of capitalist social relations. Thus as some sort of market economy

develops in Russia, historic patterns of regional growth must be taken into account.

Historic patterns of regional development

Given the physical size of the Russian Federation, it is a truism to claim that geography matters, but in attempting to understand possible patterns of change across the region, it is necessary to begin to understand the spatial legacy not only of Soviet development, but also of an earlier Tsarist phase of industrialisation. Actual geographical location also plays a role in the mediation of change, and without becoming too determinist it is essential to consider this too. Furthermore, not only is transformation leading to dramatic shifts within economic units in the Russian Federation, but it is also linked to a radical rearticulation of control over local development.

It is first necessary to understand the priority given to rapid and extensive industrialisation under Soviet rule, for this, along with the prioritising of defence and strategic industries, perhaps shapes more than anything the development of the post-Soviet space. In practice, these commitments meant that resources were concentrated in regions where rapid industrialisation could best be achieved, which were often those already developed under the tsars (Moscow, St Petersburg and the Donbass in Ukraine, for example). They were, however, set against a declared commitment to balanced and equal development (see Pallot and Shaw, 1981) which theoretically encouraged the development of peripheral and less developed regions. These often contradictory goals were played out in the Russian provinces and worked against any coherent and long-term regional development policy.

In more concrete terms, the legacies of the Soviet system which must be considered in any study of change include the extent of monopoly, the exaggerated (and regionally concentrated) military–industrial complex (*voenno-promyshlennii kompleks* or VPK), vastly distorted internal and external prices (assuming market set prices are normal), the attention paid to physical output not value or quality, and hence the wasteful use of resources, and the centrality of the Soviet-type enterprise in any locality.

Novikov argues that "the phenomenon of monopolism is rooted in the totalitarian state system" (1994, p. 1) through its systemic gigantism which meant that huge state enterprises and mega-projects came to be the norm in Soviet industrialisation. So-called monopoly producers (who, by definition, produce more than 35% of a given product) made up 18% of total Russian production in 1993. However, if you reduce this percentage to just the output of the monopoly good, the figure falls to just 7%. This small figure has been used to argue that monopolism within the Russian market is not so important. However, when the focus of attention shifts to employment and the locality, as opposed to national market shares and the question of competitiveness, concern arises over company towns, where a single large enterprise occupies a monopoly position over employment.

The issue of closed towns connects concerns over both company towns and the geographical concentration of the VPK. These towns, such as Chelyabinsk, were closed to outsiders, both Russians and foreigners, and tended to possess just one or two huge enterprises in the defence sector, on which the very existence of the city depended. In terms of interregional comparisons, in 1989, 57% of all industrial employment in Udmurtia was in defence-related industries, whereas the figure for Tuva stood at nought (IIR Inc., 1994, p. X5). There are, however, just 10 VPK enterprises in Udmurtia compared with 323 in Moscow city (Map 10.1) which suggests that such a crude figure as number of enterprises must be adjusted to account for the relative importance of defence industries in the locality. This vastly overextended defence industry is under pressure to restructure not only according to competitive influences, but also because of the end of the cold war, the subsequent collapse of the Warsaw Pact and its assured markets, and the loss of geopolitical rationale.

Focusing more specifically at the local level, the critical role played by enterprises in Soviet social and welfare provision demands attention. In the late 1980s in some cities, as much as 90% of all housing was provided for by enterprises. Industry accounted for up to 50% of investment

in schools and health. Indeed, 90% of all financial resources employed in urban areas were concentrated in the hands of industrial associations (Shomina, 1992). How local authorities and other local actors extract themselves from this situation and begin to provide for social and welfare needs independent of enterprises is an important question for post-Soviet regional development.

Andrei Treyvish, and other academics at the Russian Institute of Geography, recognise four types of region that may be "winners" in the post-Soviet transformations (Nefedova and Treyvish, 1994). These are gateway regions, natural resource regions, farming regions and, least likely of all, manufacturing regions. At least two of these are defined specifically by their "geography" – gateway regions and natural resource regions. Gateway regions are those which possess good links within Russia, and beyond to the rest of the CIS and the countries of the West. While Moscow and St Petersburg, with their highly developed infrastructures, are clear examples of this type of region, others might be Vladivostok on the Pacific coast, and Kaliningrad in the Baltic. However, it is clear that geographical location is not enough to ensure success. The city of Novosibirsk, in south-western Sibera, is frequently represented as a gateway to the east of the country, by virtue of its favourable location at the intersection of the river Ob' and the Trans-Siberian Railway and its proximity to the Central Asian states of the CIS, yet appears to be having more difficulty cashing in on this advantage.

The case of natural resource regions is perhaps more clear-cut. Those regions which record high per capita incomes are often resource-rich regions – for example, Tyumen', Sakha (Yakutia) and Magadan (Map 10.2). However, not only do income levels vary considerably within republics and *oblasts*, and between ethnic groups, but the relationship with restructuring is more complex. Resource rich Tyumen', which has an indexed per capita income of 454 (against a Russian standard of 100), records a similar level of privatisation to Ivanovo (a textile region which has a standardised per capita income of just 54 and unemployment levels of well over 20%) and Nizhny-Novgorod (a region seen to be in the vanguard of "transition" by virtue of its reformist leadership and Yavlinsky[7]-sponsored reform programme) (Maps 10.3–10.5).

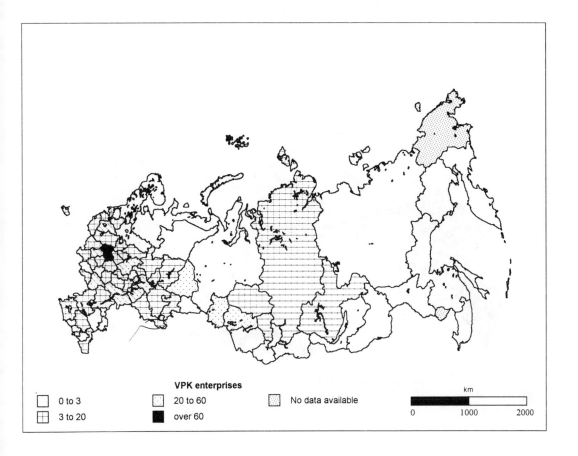

	VPK enterprises			
☐ 0 to 3	▦ 3 to 20	⬚ 20 to 60	■ over 60	▨ No data available

Map 10.1 *Number of military–industrial complex (VPK) enterprises in Russia, by oblast*

Clearly the relationship between "locational advantage", industrial geography and uneven regional development in this transformational period is not simple. It has been argued that as an economy moves towards market rationale it should be those enterprises that exhibit greatest losses that would be expected to close, and therefore cause unemployment. Thus the regions which have particular concentrations of loss-making sectors would be expected to possess highest levels of unemployment. Yet a preliminary analysis (Bradshaw, 1995) shows that industrial decline and unemployment have two quite different geographies.

To understand change in the regions then, it is perhaps necessary to look more closely at the local scale, to move away from interregional comparisons and consider what contingencies do exist in localities to mediate and complicate international and national processes of change. While "extensive" research asks the questions "what are the patterns?" and "how are the outcomes distributed?" to identify more superficial relationships, an intensive phase of study would focus on more substantial questions of causality – "what accounts for differing patterns?", "what contingencies exist in this case?", "what are (local) actors actually doing?" (Sayer, 1992, p. 243).

Shifting control over regional development

Under Soviet rule, little importance was accorded to the spatial or the local (see Chapter 7 for another perspective on local government and economic development). The sectoral, or branch, principle which afforded greatest priority to

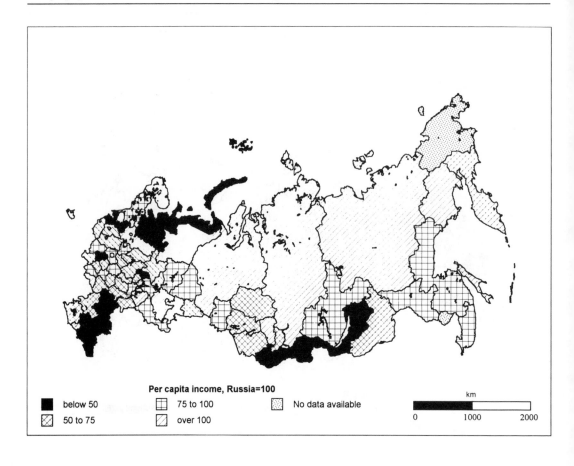

Per capita income, Russia=100

- ■ below 50
- ▨ 50 to 75
- ▦ 75 to 100
- ▨ over 100
- ▨ No data available

km
0   1000   2000

Map 10.2 *Per capita income in Russia (1993), by oblast*

intersectoral co-operation and efficiency, was set in opposition to the territorial principle which aimed to encourage more coherent regional development. Moreover, both sectoral and local issues and interests were subordinated to national economic gain. Despite a proclaimed interest in equality and balanced development, planning was "largely done for the regions, rather than by the regions themselves" (Shaw, 1986, p. 470). Regional development was theoretically very much a top-down process with regional authorities responding to central commands, and the activity of enterprises controlled by centralised ministries, and latterly, industrial associations. The great debate over Siberian development epitomised the process of regional development (see Schiffer, 1989). Even when the argument favoured the development of peripheral or underdeveloped regions, when regional development projects were benign or even beneficial, they were still directed from above. Vertical direction of the economy far exceeded any potential for horizontal co-operation. The lack of compulsory input from lesser regional authorities (*oblast*, *kray*, republic or even *raion*) in the planning process and their extreme impotence in the face of vast ministries and government departments contributed heavily to the failure of regional planning projects such as territorial production complexes. Late Soviet legislation attempted to increase the involvement of local authorities in the planning process, yet skewed attention paid to sectoral and national interests militated against any real success. Recently more attention has been paid to the fact that within the FSU there existed "a multiplicity of social relations that did not conform to officially prescribed hierarchical patterns" (Stark, 1992b, p. 300). The existence of

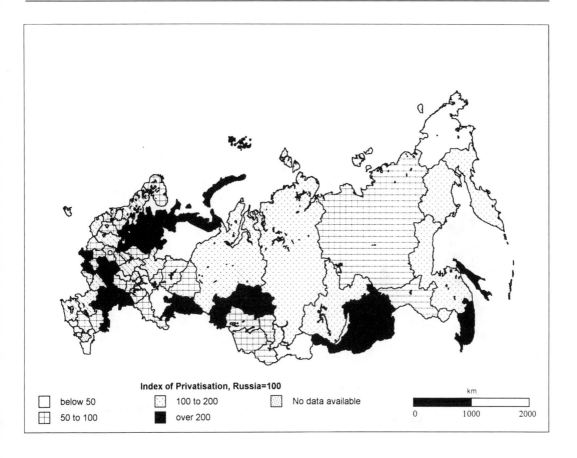

Map 10.3 *Index of all privatisation in Russia (September 1993), by oblast*

a "second economy" demonstrated the importance of relations of reciprocity and a proto-market type. Informal and inter-firm networks operated where the state could not plan, and numerous problems within the command economy meant that there were many spaces in which the informal could operate alongside the formal.

Raising questions of networks and co-operation has considerable resonance with literature on regional development in the West. The literature on the recent re-emergence of self-sustaining regional economies (see e.g. Amin and Thrift, 1992; Amin and Robins, 1990; Piore and Sabel, 1984) argues that such a re-emergence depends on an apparent (re)discovery of the importance of local linkages and co-operation and inter-firm relationships for the perpetuation of accumulation. Critics of the universal application of institutional boosterism to encourage embedded development, however, argue that such endogenous development is an opportunity only for a very small minority of localities whose historically developed structures and institutions already encourage co-operation and the collective support of common industrial agendas (Amin and Thrift, 1992, p. 585). Thus the critical question must be whether the networks and "clans" (Stark, 1992b) that existed under state socialism can provide the structural and institutional support to stimulate self-sustaining growth. The alternative is that the historical lack of coherent local development and the powerlessness of local planning bodies dooms the region to the instrumentality of international capital and the shift of control over local development from Moscow's industrial associations and centralised planning bodies to the global corporations. Endogenous development

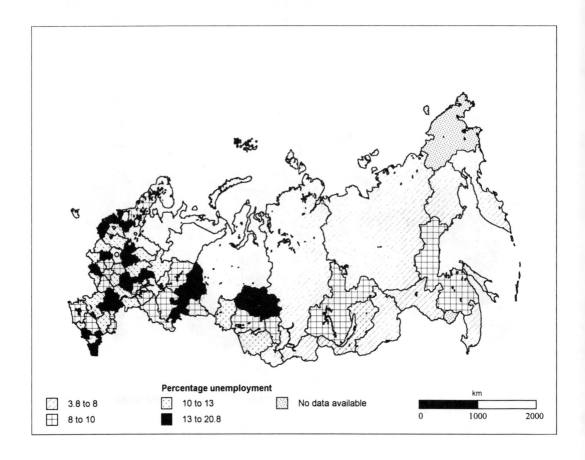

Percentage unemployment

▨	3.8 to 8	⬚	10 to 13	▨	No data available	
⊞	8 to 10	■	13 to 20.8			

km
0   1000   2000

Map 10.4   *Unemployment and short-time work in Russia (December 1993), by oblast*

and international instrumentality are clearly two extremes of global/local relationships. The reality would be better represented by a complex mix of the two, the constitution of which may be crucial to regional development potentials.

Changing social relations

One of the more formal changes in local social relations involves the emergence of financial–industrial groups (or FIGs) whose creation, it is hoped, may reconstitute former supply and co-operation linkages (see Cooper, 1995 for more details). While nationally the activity of these new FIGs is estimated to account for less than 3% of GDP, they appear to have greater significance in local areas. It has been suggested that cohesive FIGs might aid regional economic

development and permit a reduction in financial dependence on the centre. Four of the first FIGs created were indeed regional groupings, based in Izhevsk, Voronezh and Kursk, Yekaterinburg, and Novosibirsk respectively. However, more recent tendencies have shown that the predominance of local banks in FIGs is faltering as Moscow-based banks increase their involvement.

On a more informal level, it is interesting to watch the creation of looser alliances within regions that bind together a variety of local actors. Novosibirsk is an industrial region in south-western Siberia which is attempting to use this transformational moment to fulfil its potential as Russia's third largest city.[8] Academics, Western advisers, local politicians and entrepreneurs frequently suggest a number of (potentially complementary) alternatives to regional growth, including transport and distribution, hi-tech

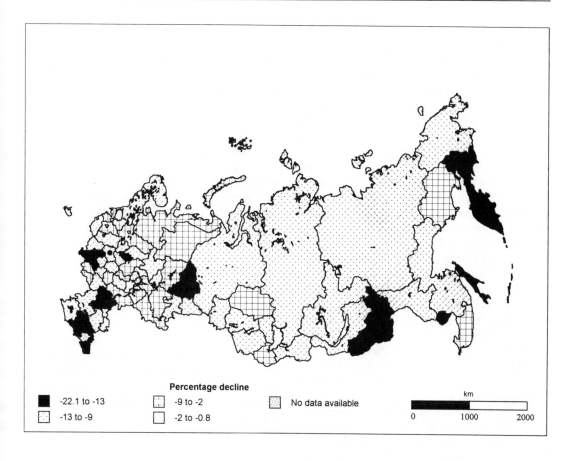

Map 10.5   *Average annual decline in industrial production in Russia 1991–93, by oblast*

specialisation and finance and administration (see Castells and Hall, 1994). The ways in which those actors cohere behind these developmental agendas may be critical to the success of Novosibirsk's transformation. In the early stages of that transformation, a number of initiatives can be seen. For example, a group of hi-tech producers and researchers have created a loose association with the aim of fulfilling Novosibirsk's much talked about scientific potential. In addition, the city now possesses a number of business associations, representing and defending the interests of various groups – small entrepreneurs and women, for example.

Perhaps as influential in the reconstitution of conditions for regional development in many Russian localities, including Novosibirsk, are foreign advisers. Novosibirsk possesses both a US AID funded American Business Centre and a small business centre, aiming to serve "as a catalyst for Russian economic revitalization".[9] These agencies encourage both incoming foreign investment and indigenous new firms. Moreover, these US agencies act with Russian and European bodies to produce promotional literature about the region, defining the region's developmental potential not only in print, but also on geographically accessible sources, such as the World Wide Web. Through many informal channels local and foreign actors are coming together to create an image of a region and to work towards the fulfilment of that image. This brief consideration of one region suggests that numerous actors are involved in the mediation of processes of reform and in the shaping of future patterns of regional development. Nevertheless, the success of these actors is open to question.

The question of regional development rests not only within local control. Despite the emergence of new arenas for control at the local level there are still other factors involved. As mentioned above, foreign actors are beginning to play a considerable part in shaping the future of the Russian regions through advice and credit. Furthermore, the level of foreign direct investment in Russia's provinces is increasing and so too therefore is foreign control over regional development. Although local control is not necessarily more benign, the increase in foreign involvement shows that the collapse of the plan will not automatically lead to local economic autonomy. In addition to foreign actors, the residual impact of national policy on local enterprises and other actors must be borne in mind. Many enterprises remain in state hands even after privatisation as a result of their strategic importance or lack of "market-value". Many more continue to rely on state subsidies and soft credits and must continue to lobby the centre for funds. Though local actors may have a certain amount of increased control, at times they must still look to the centre for assistance. To a certain extent, regional development still relies on central state benevolence, as it did during the Soviet years. The activity of apparently non-local actors mediating local, national and international change in Novosibirsk complicates any definition of "the local". It is clear that the social relations within a region are significantly conditioned by "the non-local social relations of local actors: . . . by multifarious processes occurring at wider scales – as well as by sets of localised social processes occurring elsewhere" (Cox and Mair, 1991, p. 197).

Conclusion

In looking more closely at local level change, it becomes clear that post-Soviet regional development is a complex interplay of a number of forces acting at a number of scales. Economic transformation cannot be separated from changes in social, political and cultural realms, and must be examined within its regional, national and international context. Local structures and institutions – contingencies – exist to mediate international and national processes of change and constrain or expand the choice of futures. In the Russian Federation, the interplay, in places,

of economic restructuring and spatial contingencies is an ongoing process. While the concept of market transition suggests that stasis is the norm and change simply a journey between two periods of stability, the kind of approach outlined here argues that change is constant, and that the appearance of change is determined both by "big" processes of systemic transformation and by contingent local conditions. Post-Soviet changes do vary geographically, to get beneath the "ideological and national macroscales" discussed at the beginning of this chapter and to understand the very different processes of change in the Russian Federation seems to be an urgent task.

Acknowledgements

The research of which this chapter forms a part is being carried out on an ESRC research studentship, award number R00429434312. I am grateful to Mike Bradshaw and numerous postgraduates for comments on various earlier versions of this chapter.

Notes

1. The naming of the region of study is more than a semantic issue. There are problems with many of the descriptors of this part of the world. To use "East" and "West" is to adhere to an outdated and ideologically unacceptable cold war dualism. To define the region as "post-socialist" suggests not only that those states did possess socialist political economies, but also that socialism is dead. Moreover, it confuses discussion of socialist alternatives to the neo-liberal agenda being executed in the region. "Post-Soviet" defines the rest of the region in relation to the former Soviet Union and appears to marginalise the countries of eastern and central Europe. Former Soviet Union/East Central Europe (FSU/ECE) seems to be the least problematic descriptor since it defines the region geographically and avoids the more obvious ideological connotations of the others. It does, however, shift attention away from comparisons with transforming economies beyond Eurasia – most notably China, Vietnam, Cuba and Angola, whose experiences may indeed be instructive.

2. Stark makes a point of noting that "institutions" need not be formal structures but simply "embodied routines" (Stark, 1992a, p. 21).

3. At this point "restructuring" should merely be understood as changes in the political–economic structure. The term will be more accurately defined later.

4. There were three options for privatisation open to enterprises. The second option which offers 51% of shares at July 1992 prices was the most popular, chosen by 77% of privatised enterprises by mid-1993. The first variant gave 25% of shares to the labour collective free, with options to buy further portions, and was chosen by 21%. The third option, chosen by less than 3% of firms, involves complex negotiation between parties, at the end of which the workers have the right to buy shares at nominal and discounted prices. See Slider (1994) for more details.

5. There is a vast literature on restructuring and the reconstitution of local politics which cannot be reviewed here. For an introduction see Allen and Massey (1988), Massey (1994) and Cox and Mair (1991).

6. The coherence, exclusivity and subsequent imposition of the regional development agendas of growth machines or local growth coalitions is the subject of much study. See, again, Massey (1994) on the marginalisation of women, and Cox and Mair (1991) on more general issues of coherence.

7. Grigory Yavlinsky heads one of the major non-governmental reform parties, Yabloko, which favours continued reform, but of a different nature from that of Prime Minister Chernomyrdin's Our Home is Russia (*Nash Dom Rossiya*).

8. Both Novosibirsk and Nizhny-Novgorod possess populations of approximately 1.44 million. Fluctuating populations in both these cities seem to make a definitive answer to the question of which is bigger impossible. However, as might be expected, actors in both cities will claim their city to be the larger.

9. See http://solar.rtd.utk.edu/friends/siberia/nbsc.html – Novosibirsk's business centre's pages on the World Wide Web. This is one of numerous Web pages devoted to Novosibirsk and the rest of Siberia.

References

Allen, J. and Massey, D. (eds) (1988), *The economy in question.* London: OUP.

Amin, A. (ed.) (1994), *Post-Fordism: a Reader.* Oxford: Blackwell.

Amin, A. and Robins, K. (1990), "The re-emergence of regional economies? The mythical geography of flexible accumulation", *Environment and Planning D: Society and Space,* **8**(1): 7–34.

Amin, A. and Thrift, N. (1992), "Neo-Marshallian nodes in global networks", *International Journal of Urban and Regional Research,* **16**(4): 571–87.

Blanchard, O., Dornbusch, R., Krugman, P., Layard, P. and Summers, L. (1992), *Reforms in Eastern Europe.* London: MIT Press.

Blomstrom, M. and Hettne, B. (1984), *Development Theory and Transition: the Dependency Debate and Beyond.* London: Zed Books.

Bradshaw, M.J. (1995), "Regional problems and economic transition", paper presented to the conference of the Association of American Geographers, Chicago, 14–18 March.

Burawoy, M. (1994), "The end of Sovietology and the renaissance of modernization theory", *Contemporary Sociology,* **21**(6): 774–84.

Burawoy, M. and Krotov, P. (1993), "The economic basis of Russia's political crisis", *New Left Review,* **198**: 49–69.

Castells, M. and Hall, P. (1994), *Technopoles of the World: the Making of Twenty First Century Industrial Complexes.* London: Routledge.

Chomsky, N. (1994), *World Orders, Old and New.* London: Pluto Press.

Clarke, S. (1992), "Privatisation and the development of capitalism in Russia", *New Left Review,* **196**: 3–27.

Cooke, P. (1989), *Localities: the Changing Face of Urban Britain.* London: Unwin Hyman.

Cooper, J. (1995), "Financial–industrial groups and the formation of the Russian corporate economy", unpublished paper presented at CREES Seminar on FIGs, Birmingham, 29 November.

Cox, K. and Mair, A. (1991), "From localised social structures to localities as agents", *Environment and Planning A,* **23**: 197–213.

Dolgopyatova, T. and Evseyeva, I. (1995), "The behaviour of Russian industrial enterprises under transformation", *Communist Economies and Economic Transformation,* **7**(3): 319–31.

Duncan, S. (1989), "Uneven development and the difference that space makes", *Geoforum,* **20**: 131–9.

The Economist (1995), "Russia's emerging market: a survey", 8 April.

EIU (1995, *Country Profile: Russia 1994–95.* London: EIU Ltd.

Ellman, M. (1994), "Transformation, depression and economics: some lessons", *Journal of Comparative Economics,* **19**: 1–21.

Granovetter, M. (1985), "Economic action and social structure: the problem of embeddedness", *American Journal of Sociology,* **91**(3): 481–510.

Hansor., P. (1994a), "The future of Russian economic reform", *Survival,* **36**(3): 28–45.

Hanson, P. (1994b), *Regions, Local Power and Economic Change in Russia.* London: RIIA.

Hutton, W. (1995), *The State We're In*. London: Cape.

IIR Inc. (1994), *An Industrial Atlas of the Soviet Successor States*. Houston: IIR Inc.

Kapuscinski, R. (1994), *Imperium*. London: Granta.

Kornai, J. (1992), *The Socialist System: the Political Economy of Communism*. Oxford: Clarendon.

Lovering, J. (1989), "The restructuring debate", in Peet, R. and Thrift, N. (eds), *New Models in Geography*, Vol. 1, London: Unwin Hyman, pp. 198–223.

McKinnon, R. (1991), *The Order of Economic Liberalization: Financial Control in the Transition to a Market Economy*. London: Johns Hopkins University Press.

Massey, D. (1983), "Industrial restructuring as class restructuring: production decentralization and local uniqueness", *Regional Studies*, 17: 73–89.

Massey, D. (1984), *Spatial Divisions of Labour*. London: Macmillan.

Massey, D. (1988), "What's happening to UK manufacturing?", in Allen, J. and Massey, D. (eds), *The Economy in Question*, London: OUP, pp. 45–90.

Massey, D. (1994), *Space, Place and Gender*. Cambridge: Polity.

Nefedova, T. and Treyvish, A. (1994), *Raioni Rossii i drugikh yevropeiskikh stran c perekhodnoi ekonomiki* (The Regions of Russia and Other European Countries with a Transitional Economy), Moscow: Institute of Geography.

Novikov, A. (1994), "Fenomen regional'nogo monopolizma v Rossii" (The phenomenon of regional monopolism in Russia), mimeo.

Nuti, D.M. (1994), "Mass privatisation: costs and benefits of instant capitalism", paper presented to the Economics Panel on Privatisation, BASEES Annual Conference, Fitzwilliam College, Cambridge, 26–28 March.

O'Brien, R. (1992), *Global Financial Integration: the End of Geography*. London: RIIA/Pinter.

Pallot, J. and Shaw, D.J.B. (1981), *Planning in the Soviet Union*. London: Croom Helm.

Peck, J. and Tickell, A. (1994), "Searching for a new institutional fix: the *after*-fordist crisis", in Amin, A. (ed.), *Post-Fordism: a Reader*, Oxford: Blackwell, pp. 280–314.

Piore, M. and Sabel, C. (1984), *The Second Industrial Divide*. New York: Basic Books.

Russian Economic Trends (1995), 4(1).

Sayer, A. (1992), *Method in Social Science: a Realist Approach*. London: Routledge.

Schiffer, J. (1989), *Soviet Regional Economic Policy – the East–West Debate over Pacific Siberian Development*. London: Macmillan/CREES.

Shaw, D.J.B. (1986), "Regional planning in the USSR", *Soviet Geography*, 27: 469–84.

Shomina, E.S. (1992), "Enterprises and the urban environment in the USSR", *International Journal of Urban and Regional Research*, 16: 222–33.

Slider, D. (1994), "Privatization in Russia's regions", *Post-Soviet Affairs*, 10(4): 367–96.

Stark, D. (1992a), "Path dependence and privatization strategies in east central Europe", *East European Politics and Societies*, 4: 351–92.

Stark, D. (1992b), "The great transformation? Social change in eastern Europe", *Contemporary Sociology*, 21(3): 299–304.

Taylor, P.J. (1982), "A materialist framework for political geography", *Transactions of the IBG*, NS 7: 15–34.

11

The new Central Asia: prospects for development

Ralph S. Clem
Florida International University, USA

The disintegration of the Soviet Union requires the reconceptualisation of a set of regions that together previously constituted the world's largest state. Beyond the obvious change in the sovereign status of the former Soviet republics, within all of these new countries political and economic relations are realigning, major internal social and demographic trends are now manifesting themselves, and environmental problems ranging from the nuisance to the catastrophic increasingly demand attention. Likewise, the external ties and geopolitical situation of the newly independent states are evolving, often violently, as international linkages replace what were formerly subnational ties; that is, by virtue of their new-found sovereignty the erstwhile Soviet republics are now international actors among themselves and with other states.

Nowhere within the territory of the former USSR is the need for a "new geography" more pressing than in the fledgling states of Central Asia. Partly as a legacy of the Soviet period, and partly owing to the geographical constraints and specific features of the region, Central Asia is an area under great social, economic and political stress. As the new countries attempt to maintain the levels of development achieved – at considerable human cost – during the Soviet era while at the same time implementing – to varying degrees – political and economic reforms, they face challenges arguably greater than the other newly independent post-Soviet states. Further, a redefinition of the region in geopolitical terms is under way, because both its inhabitants and other states must now confront the reality of independence.

As will be seen below, Central Asia is a region unique among other areas of the former USSR in terms of its climate and physical environment, the manner in which society evolved in relationship to the environment, the cultural, linguistic and ethnic composition of its population, and, indeed, the nature of the region's relationship to the Soviet state and to the tsarist Russian state which preceded it. Conceptually and factually, and despite its own internal differences, there is ample reason to consider Central Asia as a

separate geographical entity, both part of but very distinctive within the larger Russian/Soviet sphere. Accordingly, in this chapter we will present a basic geographic inventory of the region, including its physical–environmental, historical–political, and socio-economic characteristics, followed by an overview of the key issues and problems confronting these states collectively or individually, especially the prospects for political stability, resource development and economic growth, population change and environmental degradation.

Central Asia as a region

Regions are for the geographer a classification scheme, much as periods serve historians. As is true for chronology, there is no all-purpose definition for divisions of geographical space. For practical reasons (such as the availability of government statistical data), regional definitions often are based on political boundaries, although these boundaries usually encompass important internal differences and frequently divide like places. Thus, the operational definition of a region may not be entirely satisfactory for one's specific needs.

At the outset, we must define for our purposes the term "Central Asia", because it has been used previously to describe very different areas. Central Asia refers here collectively to five former Soviet republics which are now independent states: Kazakhstan, Kyrgyzstan, Tajikistan, Turkmenistan and Uzbekistan (see Map 11.1). As such, our usage employs the political–geographic delimitations made during the early Soviet period to define this particular territory, and therefore does not always do justice to the spatial patterns of other features of the landscape (such as the economic, physical and social). In portraying modern Central Asia, some authors have chosen to separate the northern half of Kazakhstan because, as will be discussed later, this area is unlike the remainder of the region. Further complicating matters definitionally is the fact that Soviet "Central Asia" (which in Russian was actually

Geography and Transition in the Post-Soviet Republics. Edited by M.J. Bradshaw.

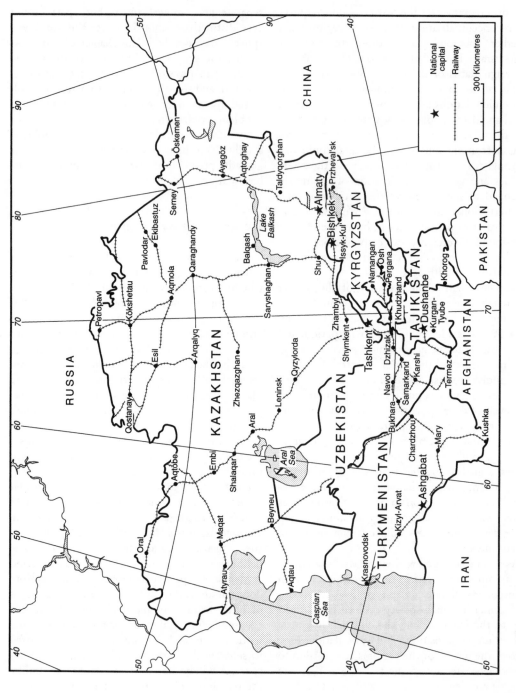

Map 11.1 *The Central Asian States*

Srednaya Aziya, or "Middle Asia") was, among other things, an economic planning region that did not include Kazakhstan. In other cases, "Central Asia" has been viewed much more expansively, stretching from the Caspian Sea eastward into western China and Mongolia ("Middle Asia" has also been used to describe this area). Finally, "Inner Asia" is sometimes seen as including at least parts of our "Central Asia", but this term is not usually meant to cover the area of the five ex-Soviet republics.

Notwithstanding the inherent problems of regionalisation, the working definition of Central Asia used in this chapter has considerable utility as a framework for describing and explaining the geographical aspect of this important and dynamic area. There is much that binds the region into a relatively coherent whole, albeit a whole with distinctive parts and, in the case of northern Kazakhstan, the potential for actual separation. The extent to which internal commonalities and differences and emerging external ties will shape the future of Central Asia are themes which will be taken up later in this chapter.

The physical environment

Stretching some 2560 km (1600 miles) from the littoral of the Caspian Sea in the west to the border with China in the east, and about 1920 km (1200 miles) from the west Siberian steppe in the north to the mountains along the Afghan border in the south, Central Asia is a vast region, incorporating an area of almost 4 million km² (approximately 1.5 million square miles). This is an area

equivalent in size to that of India and Pakistan combined. Of the countries in the region (see Table 11.1), Kazakhstan is by far the largest in area. In comparison, Kazakhstan is over twice as large as Texas, New Mexico and Arizona combined, or over 10 times the size of the United Kingdom. The smallest Central Asian country, Tajikistan, is about the size of Iowa or slightly more than twice the size of Belgium and the Netherlands together.

A comparatively inhospitable physical environment dominated by arid and semi-arid climate is one of the unifying features of Central Asia. Much of the historical genesis of the region in terms of settlement and economy is attributable directly to this fact. As will be seen below, one of the most important present-day issues confronting the region, environmental degradation, is linked to the demands placed on a limited and fragile environmental base by a rapidly growing population and a rapacious economic system.

Most of Central Asia is desert or semi-desert, created by a climate that yields little precipitation (seasonally in spring and winter) and with long, hot summers and short, cold winters. The extremes of temperature result largely from the interior or continental position of Central Asia, far removed from the moderating effects of the oceans, and the dominance of strong high barometric pressure which develops over Siberia in the winter. The lack of precipitation likewise relates to continentality and to the alignment of mountain ranges which shield the interior from moisture-laden winds.

The mountainous areas of Central Asia stretch for almost the entire length of its southern and

Table 11.1 *The Central Asian countries, 1993*

	Population ('000)	Area ('000 km²)	Population per km²	% Urban
Central Asia	52 391	3994	13.1	44.3
Kazakhstan	16 913	2717	6.2	57.1
Kyrgyzstan	4 469	199	22.5	37.2
Tajikistan	5 555*	143	38.8	30.4*
Turkmenistan	3 846*	488	7.8	44.7*
Uzbekistan	21 608	447	48.3	39.3

* Data are for 1992.

Sources: The World Bank (1994), *Statistical Handbook 1994: States of the Former USSR*, Washington, DC, p. 14. Figures for land area are from: *Narodnoye Khozyaystvo SSSR v 1990g* (1991) Moscow: Finansy i Statistika, pp. 68–72.

eastern border, and consist of several principal ranges which include some of the most impressive peaks in the world (see Map 11.2). In the extreme north-eastern corner of Kazakhstan are the Altay Mountains (elevations to 4506 m), and then in sequence to the south-west, the Tarbagatay and Dzhungarskiy Alatau straddle the border between Kazakhstan and the Xinjiang province of China. Further south, mainly in Kyrgyzstan, are the lofty (up to 7439 m) Tien Shan (Chinese for "Heavenly Mountains"). In the extreme south-east corner of Central Asia, in Tajikistan, is the Pamir range, which includes the highest peak in the former USSR ("Communism Peak", 7495 m). Finally, in the west, along the border between Turkmenistan and Iran, is the Kopet Dag range, not as high as the other ranges (maximum elevations to 2700 m).

Within and along the fringes of these mountainous areas are valleys and piedmonts which have relatively good soils and which are watered by streams rising in the higher elevations. Most notable is the Fergana Valley in Uzbekistan and Tajikistan, which is heavily populated. Two of the major rivers of Central Asia, the Amudar'ya to the south and the Syrdar'ya to the north, originate in the high mountains along the southern border and flow north-westerly to empty into the Aral Sea, while another, the Ili, has its source in the mountains of western China and flows westerly into Lake Balkhash. The Aral Sea, divided between Uzbekistan and Kazakhstan, along with Lake Balkhash and Lake Zaysan in Kazakhstan (through which flows one of the major rivers of Siberia, the Irtysh) and Lake Issyk-Kul' in Kyrgyzstan, are the largest inland bodies of water in Central Asia.

Two large deserts account for much of Central Asia's area: the Kara Kum (Turkic for "Black Sand") in Turkmenistan and the Kyzl Kum

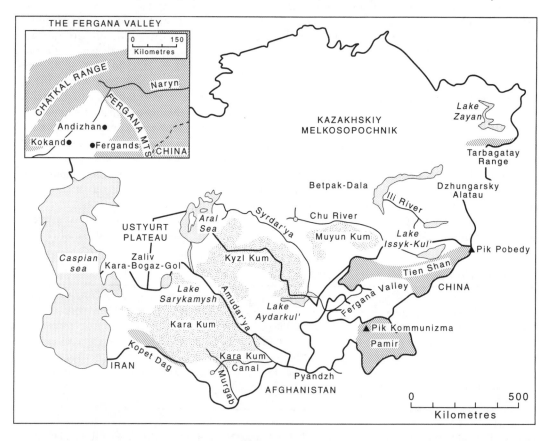

Map 11.2 *The physical setting in Central Asia*

("Red Sand") along the Uzbekistan–Kazakhstan border. The vast semi-arid interior of Kazakhstan, virtually uninhabited by sedentary population, has sparse vegetation, mainly small brush and the unique, indigenous saxaul tree. The northern border of Central Asia (i.e. the northern border of Kazakhstan) runs through the vast Eurasian steppe (prairie grassland) natural zone, with its rich chernozem ("black earth") soil.

The historical foundations of the Central Asian countries

Central Asia is an area wherein settlement dates back to antiquity; Alexander the Great passed this way, as did the fabled trade caravans of the Silk Road, and numerous invaders over the centuries subjugated the different peoples and added to the mix of inhabitants. The region has also witnessed times of great cultural, political and military influence, as evidenced in the splendid architecture of Samarkand. The most recent periods in this long history are the Russian Imperial and Soviet eras, which differ mainly in ideological terms and, as will be seen, had much in common otherwise. It is very important to understand how both of these periods influenced the economic, political and social development of today's independent Central Asian states.

The territorial expansion of the Russian state from the mid-sixteenth century on is one of the most dramatic political–geographic phenomena of all time; in a period of about 350 years, the once-humble principality of Moscow was transformed into the transcontinental Russian Empire through a process of military conquest of neighbouring lands. The advance of the Russian Empire into Central Asia, which came relatively late in this process, occurred in two phases (see Clem, 1992). First, in the eighteenth and early nineteenth centuries, the Russians moved south from fortified lines in Siberia and the southern Ural Mountains into the steppe of northern Kazakhstan, displacing or incorporating the indigenous, nomadic Kazakh people. This expanding Russian frontier was mainly agricultural, as millions of Russian, Ukrainian and other migrants poured in to cultivate the lands of the Eurasian steppe in what is today northern Kazakhstan. In the latter half of the nineteenth century, Russian troops occupied southern Kazakhstan, and from there penetrated still further south into the settled oases and piedmont, taking control of the areas presently known as Uzbekistan, Kyrgyzstan and Tajikistan. These areas of long historical settlement had, by the late nineteenth century, become relatively weak economically and militarily, and the three local states (the emirate of Bukhara and the khanates of Khiva and Kokand) were annexed by Russia or reduced to vassalage. The last part of Central Asia to be incorporated into the Russian Empire was Transcaspia (today's Turkmenistan), conquered by Russia in the 1880s after fierce resistance. The international frontiers of Russian Central Asia were delimited in 1895 by agreement with the British (whose own colonial expansion in this part of the world was northward from the Indian subcontinent). This second stage of Russian expansion into Central Asia was more in the typical European colonial style, with relatively few Russians and other outsiders dominating a much larger indigenous population.

Russian Imperial administrators organised their holdings in Central Asia mainly as governor-generalships; in this form, the territories were administered separately from provinces in European Russia, and with more power devolved to provincial authorities (Pierce, 1960, pp. 46–91). After the Soviet government established control in Central Asia, the federal structure of union republics and autonomous republics came into being (1924), whereby these units were delimited, at least in theory, on the basis of ethnic (or, in Soviet parlance, nationality) settlement patterns (Schwartz, 1992, pp. 37–73). Some analysts have concluded that the actual delimitation of these ethnic territorial units was done with care, and generally respected the distribution of ethnic groups on the ground. Yet, as Kaiser (1994, pp. 3–32) suggests, any such geographical delimitations in areas with complex settlement patterns and evolving ethnic groups will be open to criticism and will almost certainly be contested. Further, even if the attempts to define borders for the Central Asian republics were well-intentioned (and many argue that a policy of divide and rule was in fact behind these delimitations), the complexity of ethnic settlement patterns on the ground and the embryonic state of what we might call a modern ethnic consciousness mitigated against anything like ethnically pure political–administrative areas (Kaiser, 1992, pp. 284–8). Nevertheless, these republics figured prominently in Soviet life and were highly visible representations of the underlying multi-ethnic mosaic of the country's population. Symbolically, they served to legitimate not only ethnicity but, more importantly, ethnic territoriality. Ultimately, of course, these "nominally independent" republics of the Soviet period provided the perfect vehicle for exiting the Soviet Union; clearly, it is no accident that the USSR fractured along the republican fault lines. However, in Central Asia as in many other regions of the former USSR, the political–geographic legacy of the Soviet era virtually guarantees future problems with ethnic minorities and disputed borders. Whether or not these countries will continue to be viable in this form is, therefore, an open question.

Ethnodemographic composition of the Central Asian states

Central Asia is an extraordinarily diverse area in terms of the ethnic composition of its population. In addition to the five major indigenous ethnic groups of the region (Kazakhs, Kyrgyz, Tajiks, Turkmen and Uzbeks) which form the majority (or, in the case of Kazakhstan, plurality) of their respective countries' populations, there are smaller indigenous groups (Karakalpaks and Uighurs) and several important outsider groups resident throughout Central Asia (see Table 11.2).

This complex ethnodemographic composition of the newly independent countries of Central Asia derives first from the historical process whereby the Russian Empire took shape geographically and the manner in which its increasingly multi-ethnic society was formed, and then from the subsequent transformation of that huge country through large-scale economic development and social change during the Soviet era. As a result of these forces, major trends in population growth occurred in Central Asia.

Population growth is determined by the combined effects of migration and natural increase (the surplus of births over deaths). With regard to the former, interregional and rural-to-urban migrations took place into and within Central Asia, with the most noteworthy tendency being the proliferation of ethnic Russians throughout the region, especially to urban centres and to the agricultural lands of northern Kazakhstan. Ukrainians, Belarusians and Tatars likewise moved in large numbers – and in some cases, along with Germans, Koreans and other groups, were forcibly relocated – to Central Asia.

To a greater degree than other regions of the former USSR, over the last 40 years major differentials in the rates of natural increase among ethnic groups have also profoundly influenced the ethnic composition of Central Asia. The indigenous Central Asian peoples have experienced extraordinarily rapid natural increase, while the non-indigenous peoples resident in the region have not. Thus, the indigenous groups have steadily increased their share of the total population of the countries of the region, especially so in the last 10 years when the decades-long trend of in-migration to Central Asia declined and, in some cases, reversed (Table 11.2). By 1989, the date of the last census of the Soviet Union (and still the best baseline for population data), the eponymous ethnic groups of Uzbekistan and Turkmenistan accounted for over 70% of those republics' populations, while the Tajik and Kyrgyz peoples had a 62 and 52% share respectively of their republics' population (Table 11.2). In Kazakhstan, on the other hand, the Kazakhs formed a plurality, but not a majority of the population. Because demographic trends typically persist for generations, we can expect that the indigenous Central Asian groups will continue to increase their proportion of the region's population.

As noted above, a second factor determining the ethnodemographic composition of the newly independent countries was the manner in which the political–administrative structure of the Soviet Union was defined geographically. Within Central Asia, the indigenous groups are generally heavily concentrated in their respective countries (Table 11.3), but the large numbers of Uzbeks in Tajikistan and Kyrgyzstan and Tajiks in

Table 11.2 *Ethnodemographic composition of Central Asia*

	Pop ('000)		1989 as % of 1979	% of total	
	1979	1989		1959	1989
Uzbek SSR					
Uzbeks	10 569	14 142	134	62.2	71.4
Russians	1 666	1 653	99	13.5	8.3
Tajiks	595	934	157	3.8	4.7
Kazakhs	620	808	130	4.1	4.1
Tatars	531	468	88	4.9	2.4
Karakalpaks	298	412	138	2.1	2.1
Crimean Tatars	118	189	161	0.6	1.0
Kazakh SSR					
Kazakhs	5 289	6 535	124	30.0	39.7
Russians	5 991	6 228	104	42.7	37.8
Germans	900	958	106	7.1	5.8
Ukrainians	898	896	99	8.2	5.4
Uzbeks	263	332	126	1.5	2.0
Tatars	313	328	105	2.1	2.0
Uyghurs	148	185	125	0.6	1.1
Belarusians	181	183	101	1.2	1.1
Kyrgyz SSR					
Kyrgyz	1 687	2 230	132	40.5	52.4
Russians	912	917	101	30.2	21.5
Uzbeks	426	550	129	10.6	12.9
Germans	101	101	100	2.8	2.4
Kazakhs	27	37	136	1.0	0.9
Tajik SSR					
Tajiks	2 237	3 172	142	53.1	62.3
Uzbeks	873	1 198	137	23.0	23.5
Russians	395	388	98	13.3	7.6
Tatars	78	72	92	2.8	1.4
Kyrgyz	48	64	132	1.3	1.3
Turkmen SSR					
Turkmen	1 892	2 537	134	60.9	72.0
Russians	349	334	96	17.3	9.5
Uzbeks	234	317	136	8.3	9.0
Kazakhs	80	88	110	4.6	2.5
Tatars	40	39	97	2.0	1.1
Ukrainians	37	36	96	1.4	1.0

Sources: 1959 data are from Tsentral'noye Statisticheskoye Upravleniye SSSR (1962), *Itogi vsesoyuznoy perepisi naseleniya 1959 goda*, Moscow: Gosstatizdat, USSR volume; 1979 data are from TsSU SSSR (1984), *Chislennost' i sostav naselemiya SSSR*, Moscow: Finansy i Statistika; 1989 data are from Goskomstat SSSR (1991), *Natsional'nyy sostav naseleniya SSSR*, Moscow: Finansy i Statistika.

Uzbekistan are noteworthy. There are several areas, including the Fergana Valley and the area around Samarkand, that are especially ethnically diverse and therefore potentially contentious. These irredenta were of academic interest in the Soviet period for what they demonstrated about ethnosocial processes in homeland versus non-homeland situations, but in the independence era they assume much greater importance as potential sources of migration or, at the extreme, of political instability.

Of the non-indigenous, or outsider, ethnic groups of Central Asia, by the largest numerically are the Russians, who are concentrated in

Table 11.3   *Percentage distribution of Central Asian groups, 1989*

	Kazakhs	Kyrgyz	Tajiks	Turkmen	Uzbeks
Kazakhstan	80.3	0	0.6	0	2.0
Kyrgyzstan	1.5	88.2	0.8	0	3.2
Tajikistan	0.1	2.5	75.3	0.7	7.2
Turkmenistan	1.1	0	0	93.0	1.9
Uzbekistan	9.9	6.9	22.2	4.5	84.7
Central Asia	91.9	97.9	98.8	98.1	99.0

Source: Goskomstat SSSR (1991), *Natsional'nyy sostav naseleniya SSSR*, Moscow: Finansy i Statistika.

northern Kazakhstan (where locally they are often the majority of the population) and in urban and industrial areas of all the new states. This Russian dominance of most Central Asian cities and industrial concentrations has in the past served to impede the migration of the indigenous groups from rural areas to urban centres, as the indigenes found it difficult to compete for urban jobs and to deal with the Russian culture and language which prevailed there. With independence, however, has come a growing identification by the Central Asian peoples with their cities and an increasing volume of Russian out-migration, which together presage an indigenisation of urban areas in Central Asia. Other major outsider ethnic groups within Central Asia are the Ukrainians, the vast majority of whom moved to Kazakhstan (primarily to agricultural areas but also to some industrial centres); Germans, most of whom were forcibly resettled to Kazakhstan from their homeland in the Volga region during the Second World War; Koreans, who were transported to Central Asia from the Far East, mainly to Uzbekistan and southern Kazakhstan; and Tatars (of both Volga and Crimean origin). The Volga Tatars are of long standing in Central Asia, many having settled in the region in Tsarist times, but the Crimean Tatars were another of the forcibly relocated groups of the Second World War period, many of whom have recently returned to their ancestral homeland. In some cases, the presence of outsider groups has led to interethnic violence in Central Asia in the post-Soviet period, particularly in areas where economic and social conditions have deteriorated the most and where disputes over landownership have arisen. Sadly, more such cases will no doubt occur in the years ahead.

Given the importance of the ethnic basis of the new Central Asian countries, two key issues are: (1) what is the potential for further out-migration of peoples not indigenous to Central Asia, and (2) what is the potential for the migration of Central Asian peoples either to their respective homelands, to other Central Asian countries, or outside the region entirely? As regards the first question, there is a very sizeable population of non-Central Asian peoples in Central Asia (somewhere around 14 million). Clearly, nowhere near this total are likely to leave the region, as many have lived in Central Asia for decades, or are descendants of even earlier migrants. Nevertheless, as the pressures of economic decline and rapid growth of the local workforce (as discussed in a later section) combine with nationalist interests and, in some cases, violence (either directed against outsiders or merely generalised), we can expect the volume of out-migration by outsiders to increase over the next 10 years. The sources of out-migrants will probably be areas where outsiders are relatively small in numbers, areas where economic conditions are deteriorating rapidly, and areas where violence is on the increase. In Central Asia, such areas would include southern Kazakhstan, rural areas of Uzbekistan and Kyrgyzstan, and Tajikistan generally. It may also be possible that urban centres, historically dominated by Russians and other outsiders, will experience out-migration by these groups as local peoples continue to urbanise.

It is also quite likely that migrations of Central Asian peoples will increase in the direction of their respective national states, a "gathering in" process of sorts. Although information on such movements is imprecise at this writing, it is clear that Kazakhs are returning to Kazakhstan from neighbouring states (including China), and that the situation among Tajiks and Uzbeks living in each other's country is increasingly more problematic. As the pressures of economic decline and population growth increase over the coming years, it is to be expected that interethnic disharmony will rise, resulting almost certainly in a heavier volume of migration within Central Asia itself. Finally, during the Soviet period there was considerable speculation regarding the possibility of large-scale out-migration of Central Asians to other republics of the USSR, especially to those areas with chronic labour shortages and high wages (e.g. Siberia and the north). That this migration had not yet developed by the time the Soviet Union dissolved does not mean that it would not have occurred eventually, particularly in light of the steadily growing demographic pressures in Central Asia (discussed below). The independence of the Central Asian countries and the establishment of international frontiers and migration controls among these states and Russia now renders the possibility of such movements much more problematic, although not impossible. Migration ties are characteristic of most post-colonial relationships (such as the Algerian community in France or the Indian and Pakistani groups in Britain), and the eventual movement of Central Asians to Russia should not be ruled out, especially if Russia's economy rebounds.

Population growth and economic development

The disintegration of the USSR and the creation of newly independent countries from the former Soviet republics brings to the fore the complex relationships among nationalism, territory and economic development, relationships which in the Soviet era were to a large degree centrally managed and/or held in check by the pervasive one-party system and security apparatus of the old regime. A crucial factor in these relationships is population, specifically the numbers of people in different regions, their various rates of natural increase, and the migration of people from one region to another and from rural areas to cities. Through a very complicated and not always well-

understood process, regional economic development in the ethnically plural and ethnofederal USSR both created and muted interregional disparities, and took account of geographically diverse demographic trends; ironically, however, migration and natural increase often worked at cross purposes in Central Asia, as people have historically moved to this labour surplus area from labour deficit areas, especially from Russia (at least until recently). For Kazakhstan specifically, net in-migration was an important source of this growth until the 1970s, when the flow shifted to net out-migration, with the volume of out-migration increasing through the 1980s. Net in-migration to the other republics of Central Asia was never large, and in any event likewise turned to out-migration in the late 1970s.

Owing to very rapid growth since the Second World War, the countries of Central Asia now have a combined population of over 52 million people (compared with France or Turkey with approximately 56 million each). Within the region, Uzbekistan is the most populous country, followed by Kazakhstan, and then by the much smaller countries of Tajikistan, Kyrgyzstan and Turkmenistan (Table 11.1). The population of Central Asia tends to be highly concentrated spatially in those areas where irrigated agriculture is possible; that is, along the major rivers, the piedmont and the intermontane valleys, and in the agricultural steppe area of northern Kazakhstan. Despite considerable urban growth in the last several decades, and the presence of several large cities, Central Asia remains a predominantly rural region (see Table 11.1). In fact, in recent years rural population growth has outstripped urban growth, and as a consequence the level of urbanisation has gone down in all of the Central Asian states. There are, nevertheless, major urban concentrations in the mining and industrial centres of central Kazakhstan (Qaraghandy, formerly Karaganda), eastern Kazakhstan (Semey, formerly Semipalatinsk and Oskemen, formerly Ust-Kamemogrosk), and around the national capitals.

Demographically, Central Asia was a distinctive region within the USSR (Rowland, 1992, pp. 222–50). Over the last four decades of the Soviet era, Central Asia experienced the most rapid population growth of any region of the USSR, with its population increasing by more than three times. Over most of the Soviet period, tremendous declines in mortality were achieved, largely through the extension of public health and medical services. Meanwhile, fertility has remained relatively high, resulting in high rates of natural increase. As a result, the share that Central Asia accounted for of the total Soviet population almost doubled between 1951 and 1991 (from 9.7 to 17.6%). Put another way, even though Central Asia accounted for just under 10% of the Soviet Union's population in 1951, over the next 40 years it provided over 30% of population growth in the USSR. Figures for the last full year of the Soviet era (1990, see Table 11.4) illustrate the differences between the Central Asian republics, the USSR average and – for comparative purposes – Estonia in terms of birth rates, death rates, the rate of natural increase, life expectancy and the infant mortality rate. With high birth rates and natural increase, population in the Central Asian republics (save only Kazakhstan) was growing at over 2% per annum (over 3% in Tajikistan), compared with the USSR average of under 1% and under 0.2% in Estonia. On the other hand, measures of mortality which taken into account age distribution suggest that while conditions in Central Asia improved markedly since the Second World War, the region continued to lag behind other areas of the USSR. The infant mortality rate, a measure sensitive to standards of living, was appreciably higher in Central Asia than the USSR average; the rate in Turkmenistan, highest in the USSR, was almost four times that

of Estonia. Life expectancy at birth, a summary measure of mortality, likewise reflected worse conditions in Central Asia. Although fertility levels have declined somewhat in recent years, the demographic momentum of the last 30 years is such that the population of Central Asia will continue to grow rapidly well into the next century. This pattern of population change is very much in keeping with the experience of many developing countries of the world; indeed, levels of natural increase in Central Asia are generally above countries such as India and Bolivia.

Growth of the workforce in Central Asia

The major consequence of rapid population growth in Central Asia is the expansion of the workforce, which is the net growth between those joining and those leaving the working age population. One can estimate this growth by using census data by age: adding the new increments of young people and subtracting people in older ages who will be exiting the workforce through retirement gives a rough idea of how large net growth will be. For our purposes, the population aged 10–19 in 1989 equates with new entrants into the workforce for the period 1990–99, assuming that 20 is the age at which people will enter the workforce, and making no adjustments for mortality. People leaving the workforce at age 60 means that the age group 50–59 in 1989 is the number of people projected to exit the workforce over the

Table 11.4 *Demographic indicators, Central Asia, 1990*

	Birth rate	Death rate	Natural increase	Infant mortality	Life expectancy
Kazakhstan	21.7	7.7	14.0	26.4	72.6
Kyrgyzstan	29.3	7.0	22.3	30.0	72.8
Tajikistan	38.8	6.2	32.6	40.7	72.1
Turkmenistan	34.2	7.0	27.2	45.2	69.7
Uzbekistan	33.7	6.1	27.6	34.6	72.6
Estonia	14.1	12.3	1.8	12.3	74.9
USSR	16.8	10.3	6.5	21.8	73.9

Notes: Birth rate, death rate, and natural increase are per 1000 population. Infant mortality is deaths to infants up to one year of age per 1000 births. Life expectancy is number of years at birth.
Source: *Narodnoye khozyaystvo SSSR v 1990g* (1991), Moscow: Finansy i Statistika, pp. 89, 92 and 94.

same 1990–99 period. Obviously, in reality mortality will eliminate a few potential workers, and migration could either add to or subtract from the growth from natural increase, but the fact that both the younger and older age groups are already known lends credence to this procedure.

The results of workforce projections for Central Asia reveal a dramatic growth in that segment of the population. For example, in the 1990s the absolute growth in the workforce in Uzbekistan alone (1989 population: about 20 million) will exceed that of the Russian Federation (1989 population: approximately 147 million). Compared to other countries of the former USSR, the growth in the workforce as a percentage of population in the Central Asian states is much higher than in any other save Azerbaijan, and such high percentages in Central Asia portend very serious problems in so far as the absorption of these numbers is concerned.

Another demographic problem created by high rates of population growth is an unfavourable dependency ratio; that is the ratio of those in the working ages to those both younger and older (who must be supported by the working age group). In Tajikistan, for example, only 47% of the population is in the working ages, compared to 57% in Russia. Thus, whereas the working age population has grown and will continue to grow rapidly in absolute terms in Central Asia, this is compounded by a concomitant growth and a relatively high share of the total population in the dependent ages.

Economic development in Central Asia

The ability of the economies of the Central Asian countries to absorb this tremendous growth in the number of workers and to accommodate the large dependent population is questionable at best. The distribution of Central Asia's labour force is highly skewed in the direction of agriculture (see Table 11.5). Furthermore, as discussed above, the indigenous Central Asian peoples are even more concentrated in agriculture than their respective country's figures, because Russians and other outsiders typically dominate the urban-industrial/mining jobs. Beyond the usual problems faced by transitional economies (such as those described elsewhere in this volume), Central Asian countries must contend with the legacy of what amounted to a colonial, perhaps even mercantilist, relationship with Imperial Russia and later with the Soviet Union. To understand the impact of past development practice on the present and future economies of Central Asia, one should briefly survey the recent history of the agricultural and mining–industrial sectors in the region.

Agriculture

In Central Asia, the rapid increase in the rural population places a huge burden on an already overstressed agricultural economy (Craumer, 1992, pp. 132–80). At least part of the problem derives from the manner in which agriculture has been managed by Russian Imperial and Soviet authorities. Even prior to the Russian conquest, there was a significant trade between Central Asia and Russia in cotton, a crop grown in the former region since ancient times. After Central Asia was annexed by the Tsarist state, a concerted effort was undertaken by the Russian authorities to expand the cultivation of cotton in Central Asia by protecting Central Asian cotton

Table 11.5 *Distribution of workers by sector (%), 1993*

	Agriculture	Industry	Services
Kazakhstan	24.4	30.3	45.3
Kyrgyzstan	38.2	22.5	39.3
Tajikistan	44.7	20.5	34.7
Turkmenistan	44.2	20.2	35.6
Uzbekistan	43.5	21.1	35.5
Estonia	12.4	43.8	43.8

Source: The World Bank (1994), *Statistical Handbook 1994: States of the Former USSR*, Washington, DC, p. 14.

from foreign competition in the Russian domestic market, by subsidising food imports to replace grain displaced by cotton cultivation, by promoting the introduction of higher-yielding varieties of cotton and cotton-processing machinery from the United States, and by extending credit to cotton farmers. Although the result of this policy was to increase dramatically the production of cotton in Central Asia, there were also unfavourable side-effects such as the rising indebtedness of small landowners, the concentration of wealth in the hands of a very few (mostly indigenous) large landowners, and growing landlessness among the peasantry. Further, in a larger sense this policy ensured that Central Asia would be tied to Russia as a supplier of raw cotton and an importer of food and manufactured products (Pierce, 1960, pp. 163–74).

The advent of Soviet power in Central Asia brought with it the collectivisation of land and livestock, and led to severe demographic trauma in the region. The decade of the 1930s was calamitous in many areas of the Soviet Union, but no more so anywhere than in Kazakhstan. The impact of collectivisation and sedentarisation on the predominantly nomadic pastoral Kazakhs is one of the worst demographic catastrophes of the twentieth century, with losses in Kazakhstan estimated at well over 1 million persons (out of a population at the time of just over 6 million). Following this débâcle, Soviet agricultural administrators extended further the cultivation of cotton and the integration of Central Asia into the national economy, primarily as a source of raw materials. The construction of the Kara Kum canal, beginning in the mid-1950s, greatly extended the area of irrigated farming from the middle course of the Amudar'ya westward into Turkmenistan, while other irrigation projects in Uzbekistan, Tajikistan and Kyrgyzstan likewise added significantly to the cotton-growing area and to huge increases in the production of raw cotton.

Beginning in the mid-1970s, however, it was becoming clear that major problems were developing in Central Asian agriculture, particularly with cotton. As the drive to expand production necessitated the cultivation of increasingly more marginal lands, yields for cotton began to decline. Further, the demand for irrigation water grew at such a pace that outtakes from the Amudar'ya and Syrdar'ya drastically lowered the flow of these rivers in their lower courses. Perhaps worst of all, poor water management led to rising levels of soil salinity, and water returning to the rivers became highly contaminated with salt and fertiliser, pesticide and herbicide residues (Craumer, 1992, pp. 135–47).

Declining crop yields and lower labour productivity and the despoliation of vast areas through inappropriate agricultural practices, when combined with high rates of rural population growth, is a recipe for calamity in the years ahead. Unfortunately, the cost of improving the efficiency of irrigated agriculture is beyond the ability of the Central Asian countries to undertake. Making an already bad situation worse is that the over-reliance on cotton production at the expense of locally grown food meant that during the Soviet period the Central Asian republics lagged behind other areas of the USSR in measures of per capita food consumption (Craumer, 1992, pp. 147–8). Although there are some indications that the region may be shifting back to food production, major imports of foodstuffs to Central Asia will be required for the indefinite future, and at considerable cost.

Mining and industry

Central Asia is reasonably well endowed with natural resources, and the exploitation of minerals in the region goes back to tsarist times (Shabad, 1969, pp. 284–346). By the late eighteenth century, lead and silver mines were in operation in the Altay district of eastern Kazakhstan, and by the end of the nineteenth century, coal and copper were being mined in central Kazakhstan and oil was being extracted along the Caspian Sea coast of Turkmenistan, often with foreign capital. As a rule, this production was of primary raw materials only, with little processing and virtually no manufacturing accomplished locally. The Imperial Russian government also backed the construction of the three main railroad lines linking Central Asia to Russia: the Trans-Caspian line from Krasnovodsk on the Caspian Sea to Tashkent, the Orenburg–

Tashkent line, and the Turkestan–Siberia line (usually referred to as the Turk-Sib and completed after the Revolution). The railroads greatly facilitated the economic integration of Central Asia into Russia (and later the Soviet Union), serving as the medium by which raw materials (both agricultural and industrial) could be exported to European Russia. Significantly, no rail links were established between Russian/ Soviet Central Asia and neighbouring countries.

The Soviet period witnessed a major expansion of existing resource sites as well as the opening of new mining and industrial locations. Kazakhstan was the principal area for this development, especially: the coal-mining district around Qaraghandy, which became the focus of a huge iron and steel production complex begun during the Second World War; the extensive copper mining and smelting operations at Zhezqazghan (Dzhezkazgan) and Balqash (Balkhash); the iron ore deposits at Rudnyy in Qostanay (Kustanay) *oblast*, which supply the iron and steel plants in the Urals region of Russia; and the vast coal basin in northern Kazakhstan centred on Ekibastuz (the output of which is used primarily for thermal electric power generation, much of which is exported to Russia). In the post-war period, a major petroleum deposit was discovered on the Mangistau (Mangyshlak) Peninsula adjoining the Caspian Sea in western Kazakhstan, and very large natural gas fields were developed in Turkmenistan and Uzbekistan. An extensive pipeline system was constructed to transmit natural gas from the Uzbek fields north across Kazakhstan to European Russia and Siberia.

Although production of all industrial raw materials has declined since independence, the potential for future resource development in Central Asia is considerable, particularly for Kazakhstan. The leading resources for further exploitation will be petroleum and natural gas, with which the region is well endowed, and which will play a crucial role in the economic revitalisation of Central Asia (Sagers, 1994, pp. 267–98). Raw materials have the advantage of being relatively easily marketed and, because they were typically underpriced during the Soviet period, will enjoy price increases as the Central Asian economies become more integrated with world markets. Of special note is the planned expansion of the Tengiz petroleum field in western Kazakhstan, which involves foreign oil interests and which could move Kazakhstan well up the list of world oil producers. The downside to this scenario is that Central Asia's geographical location means that long hauls are required to place these products into the world market. Currently, pipeline transmission of both petroleum and natural gas must pass through Russia, and the Russian government has already used this leverage to inject itself into deals forged between Kazakhstan and Western firms. Further complicating the picture is that the Central Asian countries have little refining capacity for petroleum products, and must utilise Russian refineries to convert crude oil into fuels and lubricants. Again, the Central Asians find themselves in the neo-colonial position of raw material suppliers and finished product importers.

Even though all of the Central Asian countries save only Kazakhstan were relatively undercapitalised during the Soviet period, even those investments are drying up, and have yet to be replaced by foreign or other capital (Liebowitz, 1992, pp. 101–31). Continuing violence in Tajikistan and the persistence of *nomenklatura* regimes in other Central Asian countries discourage badly needed foreign investment, and the heretofore Russocentric view of the industrialised nations limits the amount of aid provided to Central Asia. Thus, it is unlikely that the non-agricultural sectors in the Central Asian economies, such as light industry and services, will be able to provide the hundreds of thousands of jobs needed to avoid mass unemployment and/or underemployment in the region. Large heavy industrial complexes typical of the Soviet period, such as the Qaraghandy iron and steel operation, are increasingly less relevant to the needs of the region, and the resource extraction industries require comparatively little labour.

Thus, the social and economic implications of independence are serious and largely negative. Because the possibility of expanding agriculture to accommodate the rapidly growing population is limited by the rigours of the physical environment, and especially by the scarcity of water for irrigation, the ability of the local economies to generate jobs in the non-agricultural sectors is vital. Unfortunately, it is precisely in the non-agricultural sectors that post-independence declines have been greatest. In all of the countries of Central Asia, industrial production in 1992 was down from 1991, ranging from 73% of the 1991 level in Tajikistan to 94% in Uzbekistan (Noren, 1993, pp. 424, 429). Likewise, retail sales declined in all five countries over the same period (from 28% of 1992 levels in Tajikistan to 69% of 1992 in Uzbekistan). Even more troubling for the long-term future of these economies, new fixed investment declined in all Central Asian countries save Turkmenistan. These declines portend difficulty ahead in absorbing the new entrants to the workforce in Central Asia. The scope and pace of economic reform vary widely among the Central Asian countries, with Kazakhstan and Kyrgyzstan embarking on relatively ambitious privatisation programmes, while Turkmenistan and Tajikistan are holding to what Gertrude Schroeder (1994) calls "state-managed gradualism". In the former two, privatisation of retail trade has moved ahead, and plans are in place to incorporate large industrial enterprises (with the state retaining a major share). In Kazakhstan, the state will continue to control mineral resources. In the latter three countries, privatisation is aimed mainly at small establishments, and otherwise the governments are principally concerned with reducing their dependence on cotton production by diversifying agriculture and expanding light industry.

Environmental problems in Central Asia

Like many other areas of the former USSR, the Central Asian states are faced with severe problems of environmental degradation stemming from the lack of concern for the environment on the part of Soviet officials. The overwhelming emphasis on economic growth, to the virtual exclusion of its social and ecological consequences, was a hallmark of the Soviet era generally. However, the negative impact of such unbounded development policies tends to have been greater in Central Asia, because the balance between society and the environment is very precarious in these fragile, unforgiving lands.

As noted above, the most pervasive set of environmental problems in Central Asia are those deriving from the rapid expansion of the irrigated cotton monoculture. Not only have poor agricultural techniques ruined huge areas through salinisation, but the runoff from irrigated fields returning to the region's rivers is heavily contaminated with pesticides, herbicides and fertilisers, and raw sewage is typically added to the mix. Because hundreds of thousands of people rely on major rivers such as the Amudar'ya and Syrdar'ya for drinking water, the implications of such practices is evident.

It is with the Aral Sea, however, that the consequences of the unbridled expansion of irrigated agriculture are the most severe. The region's largest body of water (and once the fourth largest in the world), the Aral Sea has been shrinking in volume and area at an alarming rate since the 1960s; by the mid-1990s, the sea was reduced in volume by over two-thirds and its area had shrunk by about one-half. This dramatic contraction in the size of the Aral Sea is due to the vastly reduced volume of water flowing into the lake from its two tributaries, the Amudar'ya and Syrdar'ya, principally because of withdrawals of irrigation water. In fact, the Syrdar'ya no longer reaches the Aral Sea, and the total inflow now into the sea is typically less than 20% of the volume in the late 1950s (Sinnott, 1992, p. 87). The effects of this wastage include the destruction of the wetlands ecosystem that formerly ringed the sea, the eradication of most fish life and the end of a once-thriving fishing industry, and the creation of a vast area of dry lake bed impregnated with the residue of the agricultural chemicals borne into the lake by river water. The desiccated lake bed has in turn been the source of clouds of toxic dust which periodically contaminate the littoral of the Ara Sea and which, combined with polluted underground water and river water, has contributed to an extraordinarily high level of morbidity among the population in the lower Amudar'ya basin. Hepatitis, dysentery, typhoid and various forms of cancer are now pervasive in the area surrounding the Aral Sea, and infant mortality has reached levels above 100 per 1000 births (i.e. very high by world standards). The prospects for improving the condition of the Aral Sea are presently not promising, mainly due to a lack of funds. Clearly, the Central Asian states, with their deteriorating economies, do not have the resources needed to reverse the decline of the sea or to put in place the programmes required to improve health conditions for the local population. Although some support from international agencies has been forthcoming, the amount to date has been only a fraction of that necessary, and one can forecast with some assurance that this ecological disaster visited upon the region will persist for at least the next decade.

Within Central Asia, there are several other environmental trouble spots worth noting. One of the worst despoiled areas is the former nuclear testing zone near Semey (Semipalatinsk) in Kazakhstan, where very high levels of residual radiation exist, and in Kyrgyzstan, the former uranium mine at Maili Sai near the Fergana Valley threatens that heavily populated area with nuclear contamination. High levels of toxic waste from non-ferrous mineral smelting are a serious problem in eastern Kazakhstan. Again, the cost of cleaning up these and other ecological hazards are daunting for the region's governments, and until they have the wherewithal to deal with the problem, the unfortunate result will be the continued high incidence of birth defects and untimely death among the populations in the affected areas.

Political stability and violence

The end of the Pax Sovietica heralded a series of violent, destabilising conflicts in several areas of the former USSR, including Central Asia. Beginning as far back as the disturbances in Almaty, capital of Kazakhstan, in late 1986, the relaxation of central authority proved deadly. Eventually, violence spread to the Fergana Valley in Uzbekistan, to rural Kyrgyzstan and most recently in Tajikistan. Thousands of refugees have taken flight from these conflicts, mostly to neighbouring homelands or, in the case of Tajikistan, to Afghanistan. Many Russians and other outsiders have left the region altogether.

The current and potential political instability in Central Asia is, as was noted above, mostly a legacy of the Soviet period. Declaring independence in 1991, as did all of the republics of the USSR, meant that the five new Central Asian states inherited the borders established by the

Soviets, as imperfect as they were. Now, with truncated local élites in power (almost all leftovers from the Soviet *nomenklatura* establishment), devastated economies and serious environmental and social problems, the fledgling states must attempt to build a viable political system, create some sense of national identity out of competing clan, local, ethnic and religious loyalties, facilitate the transition to market economies of some type, and forge new international ties (especially with Russia), and all at the same time. Not surprisingly, major difficulties have arisen throughout the region. The extent to which these difficulties are manifesting themselves in the transformation process varies widely from country to country, and warrants a brief survey of each.

Kazakhstan

The largest geographically and second most populous of the new Central Asian states, and endowed with considerable mineral wealth, Kazakhstan has been widely regarded as one of the most promising prospects for a successful post-Soviet political and economic transition. That this expectation has not materialised is testimony to the complexity of the transformation process, the negative vestiges of the Soviet era and, to some extent, the constraints of geography.

Despite its resource endowment, Kazakhstan has found it very difficult to capitalise on its mineral bounty. Domestic energy shortages plague this energy-rich country, and have seriously affected the standard of living and the functioning of economic enterprises. The principal reason for this is Kazakhstan's dependence on Russia for refined petroleum and Russia's intransigence as regards supplying natural gas and petroleum products and in curtailing the transit of Kazakh petroleum to foreign buyers (which reduces Kazakhstan's access to precious foreign exchange). Russia's manipulation of the Kazakh energy sector, which is possible because of the routeing of petroleum and natural gas pipelines from Kazakhstan through Russia, poses a threat to foreign investors (who might not be able to realise enough profit if exports are minimised by

Russia). Various alternative pipeline schemes through the Caucasus, Turkey and Iran have not materialised, and in any event all have technical or political shortcomings.

The energy problem illustrates the balancing act that Kazakhstan must perform *vis-à-vis* Russia and other external interests (i.e. China and foreign investors). Complicating this is the crucial domestic ethnic situation. As was noted earlier, Kazakhstan's population is more ethnically heterogeneous than the other Central Asian states, with the large Russian minority being especially noteworthy. Further, this ethnic Russian population is geographically concentrated in the northern *oblasts* and in the cities and industrial areas of the country. Growing dissension among the Russians (and other minority Slavic groups, such as Ukrainians and Belarusians) portends separatism in the north adjoining Russia, a cause espoused by nationalists in Russia (such as Vladimir Zhirinovskiy). Also, domestic politics in Kazakhstan have become increasingly divisive along ethnic lines. Even before the Soviet Union dissolved, ethnic Kazakhs had made dramatic political gains, and the government became increasingly "Kazakhified" (Olcott, 1993, pp. 313–30; Hyman, 1994). After the Kazakh parliament was dissolved in December 1993, new elections were held in March 1994, which resulted in an even larger majority of ethnic Kazakhs being elected to the legislature (a majority proportionally more than twice the share of Kazakhs in the electorate).

Economic reform, particularly privatisation, has also taken on ethnic overtones. Kazakhstan has embarked on a privatisation scheme wherein small enterprises are sold directly, and medium to large enterprises are privatised by the use of vouchers granted to citizens and invested through funds. Charges have been made that ethnic Kazakhs have been favoured by the government when ownership rights are apportioned. Actually, within the Kazakh group there are suspicions that clan loyalty has influenced privatisation decisions, complicating further an already difficult situation. On the ground, spatial patterns of ethnic settlement in Kazakhstan make the national government's economic reform policy even more problematic. Within any country as large and diverse as Kazakhstan, interregional

variations in resource endowment, transportation accessibility and historical investment decisions create geographical differences in the extent and form of economic development. In Kazakhstan, this has meant a heavily mining-industrialised east and centre, an agricultural north and south, and a petroleum-extractive west. Ethnic Russians dominate the industrial and mining centres of the centre and east (and the national capital, Almaty), Kazakhs the south and west, and Russians and Ukrainians the agricultural north. Thus, investment and privatisation decisions made now implicitly involve the ethnic groups resident in these areas. Granting licences to foreign firms to pursue mining or manufacturing joint ventures will most likely engage ethnic Russians, who occupy most jobs in those sectors. Likewise, the relatively underdeveloped south, which is heavily Kazakh, is desperately short of investment or even prospects. Clearly, in light of these realities, it will take an enlightened economic reform policy not to exacerbate an already tense situation, balancing Kazakh nationalism against Russian sensibilities.

Politically, Kazakhstan gives the outward appearance of stability, owing largely to the considerable skill of the only leader that independent Kazakhstan has known, President Nursultan Nazarbayev. A carry-over from the Soviet period, Nazabayev has adroitly kept Kazakh nationalists within the fold, while at the same time promoting interethnic harmony in an attempt to keep ethnic Russians more or less content. Less authoritarian than the other Central Asian states, radical nationalist groups in Kazakhstan are nevertheless harassed and disenfranchised. The dissolution of Kazakhstan's parliament in 1993 was designed to strengthen Nazarbayev's position; like many leaders of developing countries, he has sought broader executive powers to direct the reform programme without interference from a disputatious legislature. The elections of March 1994, which secured for Nazarbayev a compliant, if now relatively weak, parliament, were criticised by international monitors for a variety of election irregularities.

Kazakhstan's economy, like those of all countries of the former Soviet Union, has deteriorated steadily from independence until 1994. Gross domestic product has dropped significantly, as has its components of industrial production, electrical power generation, coal and petroleum extraction and light industry. Thousands of enterprises are idle due to mutual non-payment of accounts (enterprises owe one another but cannot settle debts), shortages of raw materials (often inputs that formerly originated in other Soviet republics are disrupted) and a shrinking domestic market. Kazakhstan's departure from the rouble zone in 1993, and the introduction of a new national currency (the tenge), caused a major monetary dislocation, and inflation has continued to hamper efforts to stabilise the financial situation. On the positive side, Kazakhstan has received considerable support from international monetary bodies, and foreign investors remain engaged if not enthusiastic.

In the months and years ahead, the key factors shaping Kazakhstan's future will be the domestic ethnic situation, the pace and scope of foreign investment (especially in the energy sector), and relations with Russia (particularly involving petroleum and natural gas). If given enough time and stability to allow for the full development of its bountiful natural resources, Kazakhstan should successfully emerge from the turmoil of the immediate post-independence period. However, if dealings with Russia and the ethnic Russian minority in Kazakhstan worsen, the potential for very serious trouble, even including the separation of parts of the country, increases dramatically.

Kyrgyzstan

A remote and mountainous country with a relatively small, ethnically diverse population, Kyrgyzstan, like Kazakhstan, faces serious difficulties as regards ethnic strife and the challenge of forging a national identity from disparate parts, while simultaneously struggling with severe economic problems (Huskey, 1993). After decades of relative numerical decline, it was only in the late 1980s that the indigenous Kyrgyz people once again achieved majority status in their republic, and even now they form only about a quarter of the country's urban population. Ethnic Russians and a large Uzbek minority dominate urban areas, and Uzbeks are

densely settled in the Osh region of Kyrgyzstan. In the waning years of the Soviet period, there occurred a resurgence of Kyrgyz nationalism in the republic, symbolised by an overtly pro-Kyrgyz language policy and the rise to power of political figures of Kyrgyz origin. Coupled with unrest in the predominantly Kyrgyz rural areas, fuelled by a rapidly growing population with few job prospects and mounting landlessness, this Kyrgyz nationalism unfortunately sparked a wave of anti-Uzbek violence which resulted in hundreds of deaths and a growing polarisation along ethnic lines.

The election of President Askar Akayev in 1991 augured well for Kyrgyzstan's transition to democratic government, if only for the fact that Akayev is the only Central Asian leader not formerly a member of the Communist political élite. Pushing an economic reform programme, Aakyev has vigorously courted foreign investment and ties with China and Pakistan. Kyrgyzstan was the first Central Asian country to introduce its own currency (the som), but rapid devaluation plunged the country into a major economic crisis by late 1993. In 1994, unable to pay for its energy imports with hard currency, Kyrgyzstan faced the cut-off of natural gas from Uzbekistan, and sharp drops in industrial production have continued apace.

Kyrgyzstan possesses substantial mineral wealth and hydroelectric power potential and has pursued the most ambitious of all privatisation schemes in Central Asia. Yet, the country's relative inaccessibility and weakly developed infrastructure mitigate against its success in transforming its economy. Although ethnic relations have quieted recently owing to the efforts of Akayev, large numbers of Russians and other non-indigenes have departed, taking with them valuable skills needed by this vulnerable state. Thus, although it has what is perhaps the most enlightened leadership in Central Asia, Kyrgyzstan faces years of uphill struggle to effect its transition to a viable state.

Tajikistan

Of the five new Central Asian countries, Tajikistan has fared by far the worst in the post-independence period (Atkin, 1993). During the Soviet era, the Tajik republic chronically lagged other areas of the USSR in general and Central Asia in particular in terms of economic development and standards of living. Attainment of higher education was the lowest of any Soviet republic, infant mortality was the second highest (behind only Turkmenistan), life expectancy was the third lowest (behind Moldova and Turkmenistan), rural population growth was the highest and the level of urbanisation the lowest of any Soviet republic. Measured in per capita terms, Tajikistan ranked at the bottom of all Soviet republics in the growth of national income and labour productivity in the late 1980s, partly due to the explosive growth of the republic's population and partly to the lack of investment there. Largely bereft of mineral resources, the Tajik economy was heavily dependent on cotton production and hydroelectric power generation.

Unique among the major Central Asian peoples, the ethnic Tajiks are linguistically and culturally Persian and of long historical standing in the region. The modern Tajik identity, however, is of relatively recent vintage, forming during the Soviet period and to a large degree as the result of the creation of a geographically based ethnic homeland. As is the case elsewhere in Central Asia as a result of the delimitation of boundaries among the Soviet republics, large numbers of Tajiks live in neighbouring countries (over one-fifth of all Tajiks reside in Uzbekistan), and the population of Tajikistan itself is ethnically fragmented, with a substantial number of Uzbeks (almost a quarter of the population) and a significant Russian minority. Complicating this scenario is a strong regional division within Tajikistan, with those from the densely settled, agricultural north around Khojand (formerly Leninabad) and centre around Kulyab at odds with the peoples of the mountainous south-west and south-east and a reform-minded intelligentsia in the capital, Dushanbe. Following independence, a government of former communists with its base in the north-centre took power (and won an election in 1994 which was widely regarded as fraudulent). A brutal civil war, in which the Russian army is heavily engaged on the government side, has been ongoing since 1992, with tens of thousands of casualties and hundreds of thousands of refugees. Peace talks between the warring factions brokered by the United Nations are scheduled for 1995, but the outlook is not promising.

The horrific implications of the civil war in Tajikistan notwithstanding, the country's economic outlook is bleak indeed. The extraordinarily high rate of population growth is producing a tremendous surplus of labour, especially in densely inhabited rural, agricultural areas. Meanwhile, the industrial and transportation infrastructure has collapsed, as evidenced by the rapid decline of production in the country's largest industrial enterprise, the huge aluminium smelter at Tursunzade. Not surprisingly, foreign investment in Tajikistan has been minimal, and assistance from international financial bodies awaits the first step towards economic reform. Even assuming that the domestic political situation stabilises, Tajikistan faces a severe test in the years ahead in restoring its economy even to pre-independence levels.

Turkmenistan

The second largest country geographically in Central Asia, but with the smallest population, Turkmenistan is a predominantly desert country with settlement concentrated in oases or clusters in the foothills of the Kopet Dag Mountains, and very sparsely populated otherwise. Cotton farming on irrigated lands along the Kara Kum canal and the Amudar'ya river provides a valuable agricultural export base, but by far the key element in Turkmenistan's economy is natural gas. Originally a by-product of petroleum extraction in western Turkmenistan, vast new natural gas fields were developed in the eastern parts of the country in the 1960s. In 1993, the present Turkmen government announced plans for a very ambitious – some analysts believe unrealistic – plan to expand both petroleum and natural gas production, drawing in Western energy firms for technical and financial support. As was noted above, however, Turkmenistan must export its natural gas through Russia, which has used increasingly tough tactics to restrict its competitors. Clearly, in relation to its population, Turkmenistan's natural resource wealth provides at least the promise of economic security in the coming years, but access to world markets remains problematic until alternative pipeline routes are realised.

Turkmenistan is also the most stable country politically in Central Asia, but only because it has not even attempted transition to a pluralist society. The president of Turkmenistan, S.A. Niyazov, who had been First Secretary of the Communist Party apparatus in the Turkmen SSR, gained his present post through an election in which he ran unopposed; in fact, no opposition political parties or informal organisations are permitted in Turkmenistan (Nissman, 1993). The heretofore relatively small declines in Turkmenistan's economy and standard of living (compared to other Central Asian countries) may be jeopardised if natural resource exports are curtailed or the energy development plan goes awry. Such an economic downturn would threaten the legitimacy of Niyazov's government and raise the possibility of anti-government agitation.

Turkmenistan also benefits from a comparatively ethnically uniform population, with the highest indigenous group percentage of the population of any Central Asian country. Uzbeks, the largest Central Asian minority, are settled mainly in the Amudar'ya river basin, and Russians, the largest minority, in the cities (especially the capital, Ashgabat, formerly Ashkhabad) and in the petroleum and natural gas extraction areas. Although quiet on the surface, clan rivalries among ethnic Turkmen persist.

Uzbekistan

The most populous and densely settled of all the Central Asian countries, Uzbekistan stretches from the rich agricultural Fergana Valley in the east to the shores of the Aral Sea in the west. The largest city in Central Asia, Tashkent, is the national capital, and several famous cities of Central Asian antiquity, such as Bukhara, Khiva and Samarkand, are within its borders. With large natural gas deposits and cotton production, Uzbekistan is nevertheless beset with a huge labour surplus (resulting from high rates of population growth) and massive unemployment. Economic reforms announced with much fanfare

in 1994, to include more rapid privatisation and laws to protect private property rights, have yet to have much effect. Although the decline of Uzbekistan's economy has not been as steep as the other Central Asian countries, it none the less urgently needs foreign investment, especially in the manufacturing sector, to create jobs for the burgeoning labour force.

Uzbekistan's political climate is such, however, that foreign capital has been slow to enter the country (Gleason, 1993). Uzbek president Islam Karimov, a former communist official who transitioned in power from the Soviet period, has put in place an authoritarian government and almost eliminated his political opposition. Nationalist groups established during *perestroika–glasnost* (namely, Birlik and Erk) have been driven underground or otherwise neutralised, while leaders of human rights organisations are regularly assaulted. Elections to the Uzbek national legislature in December 1994, not surprisingly were dominated by Karimov's ruling Democratic Party, as virtually all other interests were denied places on the ballot. This tendency towards political intolerance and a pervasive climate of corruption and nepotism antedates independence; in fact, perhaps the most notorious embezzlement scheme of the Soviet era involved numerous high-ranking officials and government and party functionaries in Uzbekistan with ties to the cotton industry.

External relations

Since independence, the Central Asian countries are international actors in their own right, but remain subject to strong geopolitical constraints. Most importantly, Russia still looms very large on the Central Asian horizon for at least three crucial reasons. First, Russia currently controls virtually all land transportation access from Central Asia to the rest of the world, especially the rail and pipeline links required to export mineral resources and other goods (such as cotton). Although work is already under way to connect the Trans-Caspian line in Turkmenistan with the Iranian rail network (at Mashhad in north-eastern Iran), which would provide access to the Iranian Persian Gulf port of Bandar Abbas, and to upgrade roads from Afghanistan and Pakistan north to Turkmenistan, Uzbekistan and Kyrgyzstan (in the latter case, the famous Karakoram highway through China), these routes are difficult and not reliable enough to carry the full volume of trade necessary to sustain the region. Likewise, a new rail link from Kazakhstan east through the Alatau pass into the Xinjiang province of China, and thence to eastern China and coastal ports is open, but has yet to carry heavy volumes of freight. Alternative pipeline routes are also being discussed, some of which would bypass Russia altogether, but as was noted above, there are technical and political difficulties with each option. In the mean time, Russia has used its control over access to leverage its influence in Central Asia, and will no doubt continue to do so.

The large ethnic Russian presence in Central Asia is the second factor at play in the relations between Russia and the region. Russian nationalist politicians and the Russian military leadership have stated clearly their commitment to their compatriots in the so-called "near abroad", and it is easy to imagine scenarios wherein Russian intervention would be possible if growing instability threatened violence against Russians in Central Asia. Russia views its security interests in Central Asia as vital, hence its commitment of over 25 000 troops to maintain relative peace in Tajikistan. Finally, Russia remains the leading trading partner in terms both of imports and exports for all Central Asian countries (save only exports from Turkmenistan, for which natural gas supplied to Ukraine tops the list). This pattern can be expected to continue indefinitely for reasons of geographic proximity and the inertia typical of post-colonial relationships, reinforced in this case by the Russocentric alignment of land transportation routes.

In addition to Russia, China, Turkey and Iran have become increasingly more engaged in the new states of Central Asia. All have signed economic and trade agreements with various Central Asian states, and the Iranian-sponsored Economic Co-operation Organization, sometimes referred to as the "Islamic Common Market", has at least the prospect of enhancing linkages among its members (Iran, Turkey, Pakistan, Afghanistan, Azerbaijan and the five Central Asian countries). South Korea has shown an increasing interest in the region, especially Kazakhstan and Uzbekistan (with their large Korean ethnic minorities). Air routes have opened up from a number of foreign countries to destinations in Central Asia, including service from Europe to Tashkent and Almaty.

The international borders of Central Asia present some irredentist problems, which have to date not progressed beyond the talking state but which – just as in the case for the imperfectly delimited borders among the Central Asian states themselves – suggest the possibility of future difficulties. Most significantly, there are large numbers of ethnic Turkmen in Iran and Afghanistan, a large Tajik population in Afghanistan, and a substantial Kazakh population in western China. In the post-independence period, contacts between these groups on either side of the international frontier have increased, and some migration has taken place from surrounding countries to Central Asia. As the Central Asian states mature, it is not out of the question that demands for even more open borders will be heard.

Conclusions

A combination of demographic, ethnic, economic and environmental concerns portend serious problems in the years ahead for the newly independent countries of Central Asia. Faced with a rapidly expanding workforce, an agricultural resource base already strained by the pressures of rural population growth and an increasingly degraded environment, governments in the region will be hard pressed to prevent a further deterioration in living standards, and ultimately might face challenges from a disenchanted populace. Relations with Russia are vital for these states, because Russia exerts much influence on the development of the mineral resources which hold the key to improvements in the economies of Central Asia. In the years and decades ahead, Central Asia is likely to be a region of stress and unsettled political conditions, as the peoples and governments of the region attempt to deal with the legacy of Russian colonialism and Soviet development.

References

Atkin, M. (1993), "Tajikistan: ancient heritage, new politics", in Bremmer, I. and Taras, R. (eds), *Nations and Politics in the Soviet Successor States*, Cambridge: Cambridge University Press, pp. 361–83.

Clem, R.S. (1992), "The frontier and colonialism in Russian and Soviet Central Asia", in Lewis, R.A. (ed.), *Geographic Perspectives on Soviet Central Asia*, London: Routledge, pp. 19–36.

Craumer, P.R. (1992), "Agricultural change, labor supply, and rural out-migration in Soviet Central Asia", in Lewis, R.A. (ed.), *Geographic Perspectives on Soviet Central Asia*, London: Routledge, pp. 132–80.

Gleason, G. (1993), "Uzbekistan: from statehood to nationhood?", in Bremmer, I. and Taras, R. (eds), *Nations and Politics in the Soviet Successor States*, Cambridge: Cambridge University Press, pp. 331–60.

Huskey, G. (1993), "Kyrgyzstan: the politics of demographic and economic frustration", in Bremmer, I. and Taras, R. (eds), *Nations and Politics in the Soviet Successor States*, Cambridge: Cambridge University Press, pp. 398–418.

Hyman, Anthony (1994), *Political Change in Post-Soviet Central Asia*. London: Royal Institute of International Affairs.

Kaiser, R.J. (1992), "Nations and homelands in Soviet Central Asia", in Lewis, R.A. (ed.), *Geographic Perspectives on Soviet Central Asia*, London: Routledge, pp. 279–312.

Kaiser, R.J. (1994), *The Geography of Nationalism in Russia and the USSR*. Princeton: Princeton University Press.

Liebowitz, R.D. (1992), "Soviet geographical imbalances and Soviet Central Asia", in Lewis, Robert A. (ed.), *Geographic Perspectives on Soviet Central Asia*, London: Routledge, pp. 101–31.

Nissman, David (1993), "Turkmenistan: searching for a national identity", in Bremmer, I. and Taras, R. (eds), *Nations and Politics in the Soviet Successor States*, Cambridge: Cambridge University Press, pp. 384–97.

Noren, J.H. (1993), "The FSU economies: first year of transition", *Post-Soviet Geography*, **34**, September: 419–52.

Olcott, M.B. (1993), "Kazakhstan: a republic of minorities", in Bremmer, I. and Taras, R. (eds), *Nations and Politics in the Soviet Successor States*, Cambridge: Cambridge University Press, pp. 313–30.

Pierce, R.A. (1960), *Russian Central Asia: 1867–1917*. Berkeley: University of California Press.

Rowland, R.H. (1992), "Demographic trends in Soviet Central Asia and southern Kazakhstan", in Lewis, R.A. (ed.), *Geographic Perspectives on Soviet Central Asia*, London: Routledge, pp. 222–78.

Sagers, M.J. (1994), "The oil industry in the southern tier former Soviet republics", *Post-Soviet Geography*, **35**, May: 267–98.

Schroeder, Gertrude E. (1994), "Observations on economic reform in the successor states", *Post-Soviet Geography*, **35**, January: 1–12.

Schwartz, L. (1992), "The political geography of Soviet Central Asia: integrating the Central Asian frontier", in Lewis, R.A. (ed.), *Geographic Perspectives on Soviet Central Asia*, London: Routledge, pp. 37–73.

Shabad, T. (1969), *Basic Industrial Resources of the USSR*, New York: Columbia University Press.

Sinnott, P. (1992), "The physical geography of Soviet Central Asia and the Aral Sea problem", in Lewis, R.A. (ed.), *Geographic Perspectives on Soviet Central Asia*, London: Routledge, pp. 74–97.

12
The Russian Far East: Russia's gateway to the Pacific

Peter de Souza
University of Gothenburg, Sweden

Introduction

A central point of departure for this chapter is the fact that the Asia–Pacific region more and more appears as an important, or *the* important future, centre for the world economy. A region where a concentration of international co-operation and competition takes place and where a multitude of regional and national identities interact in economic, social, cultural and geopolitical dimensions. Indicative of this fact is that the share of the Asia–Pacific region in world trade grew from around 30% to almost 40% between 1965 and 1989.

An aspect of this important general tendency, that has not been studied and analysed enough, is the Russian presence. Within contemporary global development tendencies, the wave of reform that permeates Russia and the obvious future linkages make the Russian Far East (FER) and, in an even wider geographical context, Siberia and the Russian Federation, an important dimension of Pacific economic and political development. Kalinin in 1923, Zhdanov in 1939, Mikoyan in 1945, Khrushchev in 1954 and 1959, Brezhnev in 1966 and 1978, Gorbachev in 1986 and 1988, all spoke about a new era of engagement by the Soviet Union in the Pacific and argued in terms of regional development, economic co-operation and an intensified ambition for peace. It seems that the time has come when a Pacific orientation will become a reality. The population, resources and ports of the FER are now more important to Moscow and Russia than they have ever been before. The assertiveness of the Eurasian strain in Russian foreign policy is becoming increasingly prominent.

The problem

The use of the symbol of a gateway in the title indicates the complexity of the analysis and the variance of potential outcomes. To take this metaphor further, a gate can be closed or open, it can be large or narrow, it can be the first in a line of gates before you reach the open, it can be controlled by a security guard and need special permits, etc. The metaphor is only valid as long as it symbolically enumerates and indicates the extent and substance in the existing and future processes of Russian international relations and the potential role of the FER in this.

The objectives of this chapter are:

1. To describe, in general terms, the potential of the Russian FER in terms of raw materials, energy sources and industrial structure;
2. To present the institutional, structural and economic factors and processes that hinder an accelerated economic development of the region;
3. To discuss the dimension of the Russian Federation and its presence in the region, especially the FER in the Pacific region.

The geographical context

A number of delimitations of the geographical area could be presented. As the primary concern is the economic reorientation of Russia and its FER, including primarily political consequences, a geographical delimitation should be made. The reader should at the same time recognise the continuity of economic and political variables, which could be presented with their roots all the way to the European parts of Russia. As a valid example of this stands the reduction in port facilities connecting Russia with the open seas. The ports in Nakhodka, Vostochny and Vladivostok have all increased substantially in importance, as evidenced by increased volume of goods transferred. Still, it is possible to discern a breakpoint somewhere in the vicinity of Lake Baykal, defined by a break-even point of gas transport, and arguments raised in this chapter, in their general conclusions, are valid for this territory. The focus on this chapter is, however, directed to what has been and is defined as the far eastern economic region and its constituent areas (see Maps 12.1 and 12.2).

The FER encompasses an area of approximately 6 million km². This accounts roughly for

Geography and Transition in the Post-Soviet Republics. Edited by M.J. Bradshaw.

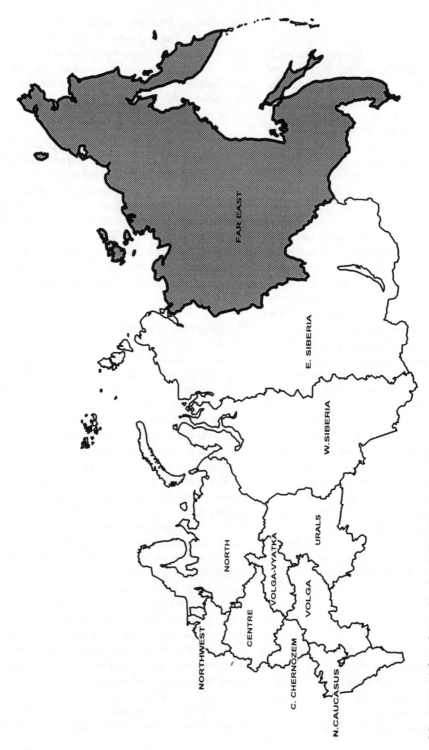

FAR EAST

E. SIBERIA

W. SIBERIA

URALS

NORTH

VOLGA-VYATKA

VOLGA

CENTRE

C. CHERNOZEM

N. CAUCASUS

NORTHWEST

Map 12.1   *The relative location of the Russian Far East*

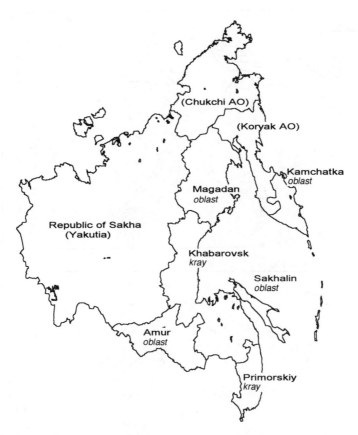

(Chukchi AO)

(Koryak AO)

Kamchatka
oblast

Magadan
oblast

Republic of Sakha
(Yakutia)

Khabarovsk
kray

Sakhalin
oblast

Amur
oblast

Primorskiy
kray

Map 12.2 *The administrative structure of the Russian Far East*

about 36.5% of the territory of the Russian Federation. Apart from the internal, administrative delimitations, the FER borders the People's Republic of China and the People's Republic of Korea and is washed by the Arctic and Pacific oceans. Separated by narrow waters, the FER also borders on Japan and the USA (Alaska). The FER region covers a north–south extent of approximately 4800 km and west–east of 4000 km. The latitudinal extent ranges from the Arctic to a southern tip at approximately the same latitude as Rome and Madrid. The region is extremely varied in physical terms and to generalise about the climate or the physical geographical features in a few words is not really possible (Map 12.3a–c). The first aspect of it is, naturally, the landmass as such.

Large areas find themselves far from the moderating influences of the world's oceans. The northern latitude of the most important waters,

the Arctic and Pacific oceans, which makes them frozen a substantial part of the year, little diminish the extreme continentality of the climate. Mountainous terrain along the eastern and southern boundaries cuts off any moderating oceanic influences. The picture of extreme continentality is complemented by a coastal climate. Far Eastern temperatures vary considerably, from the northern parts of the Republic of Sakha January temperature goes down to –50°C, compared with the same time of year in the Kurile Islands where it ranges around –10°C. Summer temperatures have a more even spread from 9 to 21°C (*Dalniy Vostok Rossii – Ekonomisheskoye Obozrenie*, 1993, p. 18). Year-round temperatures are lower than in other parts of the Russian Federation at equivalent latitudes. The length and coldness of the winters are one of the main factors limiting economic activity (Kovrigin, 1986, p. 7). Almost the entire mainland part of the FER lies within

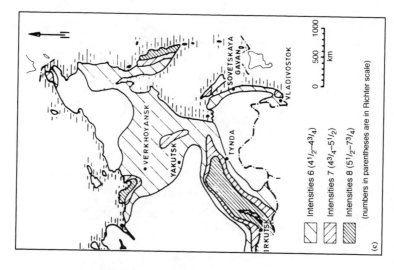

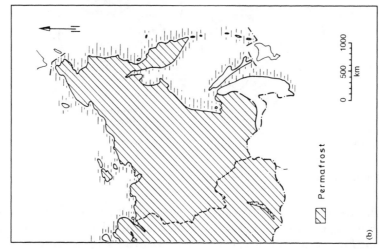

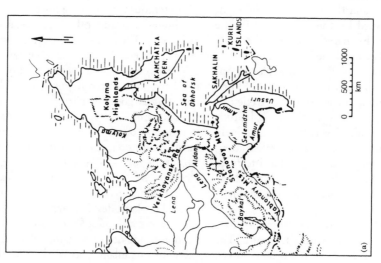

Map 12.3 *(a) Physical features of the Russian Far East; (b) areas affected by permafrost in the Russian Far East; (c) seismic activity in the Russian Far East*

the permafrost zone, which makes construction work more difficult, maintenance of roads and railroads more laborious and costly and agriculture activity troublesome, if not impossible. Seismic areas of importance appear around Lake Baykal and north-eastwards, Kamchatka (especially the eastern part) and the Kurile Islands chain (Mote, 1983).

The region is sparsely populated with a total of 8.057 million inhabitants[1] or about 5.4% of the Russian Federation. For an administrative disaggregation see Table 12.1. This constitutes a population density for the FER of 1.3 inhabitants per km^2, which is also disaggregated in the table. This table points out the wide variations. Three-quarters of the population is found in the southern parts of the region (an area which accounts for about 10% of the area of the FER) and half live within 80 km of the sea (Rodgers, 1990, p. 1). The level of urbanisation is over 75%. At the time of the 1989 census, the ethnical composition was 79.8% Russians and 7.8% Ukrainians. The share of other nationalities was substantially lower, for instance 3.6% Yakutians, 1.6% Belarusians and 1.1% Tatars and peoples of the northern territories (*Dalniy Vostok Rossii – Ekonomicheskoye Obozrenie*, 1993, pp. 26–30).

Historical legacy

Runaway serfs and fur hunters were the first to penetrate the Ural Mountains in search of freedom or riches. Soviet/Russian historians usually claim that the first organised efforts of penetration, leading up to the final conquest of Siberia, started in 1581, when the cossack Yermak set out for the East. Ivan IV gave his consent and the wealthy merchant family Stroganov financed the effort. They were given large concessions for their support. Yermak was followed by a wave of fortune-hunters in quest; primarily, for the "brown gold" (furs), which made a substantial contribution to the Tsar's coffers and the hunters and merchants involved (see e.g. *Russia's Conquest of Siberia*, 1985). The penetration reached the Pacific coast in 1639 and the first stronghold, Okhotsk, was built in 1649. The wave of fortune-hunters and the interests of the Russian Empire came, however, to collide with Chinese interests. China was, at this point in time, too strong to grapple with, and relations were regulated in the Treaty of Nerchinsk in 1689. Instead the Russian interests turned in a north-easterly direction. Fur hunters and merchants explored the northern reaches of the continent, turning their attention to the islands in the Bering Sea and finally passed the Bering Strait over to Alaska. Russian interests remained high over a long period of time and reached far south along the American coastline establishing a short-lived colony near San Francisco (see e.g. Starr, 1987). The middle of the nineteenth century saw a period of substantial change. Increased attention from English merchants and fur-hunting interests in Alaska and further penetration from British interests in the vicinity of Russia's southern borders, forced the Russian Empire to sell its interests in the American continent for 7.2 million dollars in 1867 to the USA. This coincided with an increased emphasis, from the Russian side, on the areas just north of the Amur river and the river itself.

Table 12.1   *Territory and population of the Far East economic region administrative areas*

Administrative unit	Area ('000 km^2)	Population ('000)	Population		Population per km^2
			Urban	Rural	
Republic of Sakha/Yakutia	3103.2	1108.6	737.9	370.7	0.4
Primorskiy *kray*	165.9	2299.6	1784.6	515.0	13.9
Khabarovsk *kray*	824.6	1850.7	1461.6	389.1	2.2
Amursk *oblast*	363.7	1073.7	729.5	344.2	3.0
Magadan *oblast*	1199.1	533.7	435.7	98.0	0.4
Sakhalin *oblast*	87.1	717.5	611.9	105.6	8.2
Kamchatka *oblast*	472.3	472.8	384.4	88.4	1.0

Source: *Investor's Guide* (1993, p. 58).

Internal strife in the Chinese Empire reduced the ability for it to defend its regional interests. It was forced to concede to Russian territorial ambitions in the treaties of Argun in 1858, and Peking in 1860. The nature of the Russian presence in the region is of interest, as it is not until the end of the nineteenth century that the Russian state develops a more comprehensive economic and political approach. It is in the 1880s and 1890s we can discern an ambition for populating the region, where subsidising settlers and promoting settlement schemes in general comes to the forefront. A central role in this is taken by the construction of the Trans-Siberian Railway. The patterns of settlement had, of course, an intimate relationship with the dimension of physical geography covered in the preceding section.

Having subdued the Chinese, a new contender for regional control appeared on the agenda, and that was Japan. The outbreak of the Russo-Japanese War put an end to unhindered Russian expansion, and also presented the first signs of what was to become the end of the Tsarist period. The October revolution in 1917, the civil war and intervention characterised the first years of Soviet power. In Siberia and the Far East, it was primarily the white generals Kolchak and Pepelyev that represented the anti-Bolshevik forces. It was not until 1923 that the Red Army defeated the last remnants of the white forces. The new state had, under pressing economic circumstances, little interest in the eastern parts of the country. Economic necessities made a focus on the European areas self-evident. The awareness of tremendous natural resources with the Asian territory was, however, present.

The First Five-Year Plan did not, to any great extent, show an opening towards the east, although the Ural–Kuznetsk project came to play a role as a bridgehead and a model for coming decades (Holzman, 1957). The 1930s also saw the first projections and debate over the Angara project, i.e. the harnessing of the River Angara for its energy potential. This latter project was important, on its own, as it linked the issue of relatively cheap energy with future industrial development in east Siberia, a recurring theme in later decades. The project was not started, however, until the end of the 1950s. Although starting out already directly after October 1917, but in substance in the early 1920s, it is not until the end of the decade that we can see the basics of the "Soviet system" appear. Institution building, in the widest sense of the word, started out in this period. Organisational work focused especially on the introduction of the Five-Year Plans as the main development instrument, and the special economic priorities that were to prove their worth in the 1930s and 1940s, form a cornerstone in the understanding of economic and political structures and processes of later decades. A basic theme in this development, which will be returned to, was the centre–peripheral structural relations where Moscow became the focal point. A centre dominating the connected regions in a radial network and where any eventual intra-peripheral relations were scorned upon or effectively controlled by the centre.

An important turn of priorities came about in connection with the Second World War, when the approach of German forces and later occupation, meant that already developed plans for relocation of industrial enterprises to west and east Siberia changed the economic–geographical profile of the country in a substantial way. Wartime investments further increased this tendency. Although focusing on economic priorities, it would not be correct to delimit the Soviet priorities in this sphere alone. The Soviet Union did, through Comintern, play a very active role in forwarding its own position. It did this in the name of the international proletariat and with an active "support" for a large number of different movements and activities. The Comintern activities in China should be noted as an important case in point. With the late entry of the Soviet Union in the war against Japan, and in connection with the repercussions of the agreements in Tehran, Potsdam and Yalta, and with the direct and indirect territorial demands made by Josef Stalin, there was a marked change in priorities within the international community. The emergence of the cold war and the status of the Soviet Union as a superpower, demanded a political, economic and military presence in the Pacific region and strategies for the development of these ambitions. In the beginning of the *perestroika* period the Soviet forces in the region or connected with it were considerable. Ground forces: 56 divisions (500 000), air force: 2390 combat aircraft, naval forces: 840 ships and a considerable nuclear potential (Weiss, 1989, p. 15). During the 1970s and early 1980s a changed political strategy in the Pacific area, from a military strategic and security policy emphasis to an economic orientation, can be identified. That Siberia and the FER would successively play a more and more vital role for the international economy, became evident. That the international market and economy would also play a vital role for the Siberian and Far Eastern economy slowly began to be understood. That this would have important implications for raw materials, energy and trade policies and connected with those, geopolitical and security policy considerations, was also logical.

The status of the economic situation and economic development during the Brezhnev period did not, however, give much room for either an intensive or an extensive development of Siberia and the FER. Instead the economy turned into a free-fall that, in spite of all reform efforts, led to Mikhail Gorbachev's ascendancy, and the introduction of *perestroika*. The process of *perestroika* and *glasnost* did open up a Pandora's box of political and other issues, which forced any kind of strategic thinking and policy-making to the background. In the *perestroika* period, efforts to launch a major economic restructuring programme for the FER (1 April 1989) were made. Due to the general political situation, the programme lacked realism and met bureaucratic obstruction. It did not become more than wishful thinking. Economic and political realities, still prevailing, showed a clear focus on Russian–European priorities and very short-term necessities.

The Yeltsin administration has also (*Commersant*, 6.10.92), through a decree, showed its ambition and taken a stand behind a concentrated effort on the FER and the importance of the west Siberian oil and gas industry. What remains to be seen, and where the latest years have given a disconcerting impression, is the enormous difference that exists between words and deeds. The acute economic and political situation that prevails in the Russian Federation in 1994–95 does not give any real possibility of long-term investment strategies. The reality of the present situation is that the administrative regions of the FER cannot be sure of central allocations of investment and any other resources; in this case it is natural that the periphery will seek other kinds of solutions.

The political and economic turmoil

The development of the regions has its roots in the Imperial administrative system. On top of this geographical pattern the command–administrative economic system has imposed structural features, among which a high degree of specialisation stands out in importance. Problems of weak or negative growth and insufficient or misdirected investment resources are also pertinent. Failure to reach decided or defined production targets and problems with deliveries (timing, quality and assortment) occur, as do weak development of productivity and bad co-ordination of production and between production and infrastructure and, in general, bad quality of production. These are all factors describing the workings of the Soviet-type economy of the 1960s up until the 1980s.

Some of the structural features of the old economic system and its general shortcomings are still evident in the economy of present-day Russia. The legacies of this old system remain in many dimensions and serve to constrain policies aimed at creating a new and functioning one. The period since the start-up of *perestroika* and especially since the formation of the Russian Federation has been, and still is, characterised by large budget deficits and an intensive inflationary process sometimes riding on the brink of a hyperstage (with rapid wage increases and reductions in real incomes).[2] There are large-scale declines in industrial production with resultant negative growth and prospective redundancies and incomplete institutional development and incomplete and inconsistent legislation. The breakup of the union and consequently the dissolution of traditional linkages in the economy have hindered the process of restructuring and vice versa.

The development has, in its territorial dimension, also been determined by simultaneous regional and local initiatives and processes, as distinct from central. In a period characterised by severe economic problems, decision-makers on all levels and directions tend to focus efforts and resources under their control on short-term solutions. This has an important geographical effect, i.e. a preference for cost-efficient, profit-oriented locations and activities. Considering the extremes of the present economic situation it could be argued that the predominance of short-term measures will put new economic problems on the agenda in the not too distant future. This assumption is fundamental for the scenario perceived in this case. In a more positive economic climate there will be a tendency for both state and private economic actors to invest and/or allocate absolutely and relatively more in some of the eastern regions. This is in spite of the fact that the expectations of return on investments will be more long term and the yield less than might be obtained from similar investments in places closer to the economic centre of Moscow. The last proposition can seem doubtful but is based on the assumption that comprehensive, long-term economic development is a feature of, and a basis for, long-term economic growth, and is recognised as such by the state authorities.

The economy of the Far Eastern economic region

Summarising the FER economy, it can be described as inefficient, unsophisticated and with a narrow production profile. It is largely dependent upon raw material and energy extraction. Another aspect of this profile is a generally low processing capacity. The non-ferrous metals industry together with fishing and forestry constitute the core of the economy (more than 50% of the region's industrial production).[3] Fishing alone accounts for one-quarter of the total. A very high dependence on the defence industry and military-related activities is also significant, due to the general lack of industrial activity. The region has a poorly developed infrastructure. The following will describe this profile in more detail, but first let us consider the region's advantages.

In the middle of the eighteenth century, Lomonosov, the great Russian scientist, stated that in Siberia rests the source for Russia's status as a great power in economic, political and military terms. In the beginning of the 1990s the FER produced 30% of the timber in Russia, 7.3% of the natural gas, 5.5% of the iron ore and 5.2% of the coal, while its manufacturing industries all in all accounted for 29% of Russia's total marketable output (Russia's interests . . ., 1995, pp. 20–1).

In general, the energy resource reserves in the FER are enormous (Map 12.4). The FER was estimated to have about 30% of the Soviet coal. Proven reserves are almost 13 billion tonnes, with probable reserves 15 times this. Half of this can be mined in open pits (*Forbes*, 26.11.90). Oil, natural gas deposits and hydropower sites are abundant and widely dispersed. The FER is also described as one of the richest forest regions of the Russian Federation. It has 507 million ha of land in the state forest reserve, and 351 million ha of forest land, of which 270 million ha is forested; 26 million ha is covered by brush and a further 16% is classified as protected (Barr, 1990, p. 121). The total volume of growing stock is 21.3 billion m³, of which 13 billion m³ comprises conifers.

The FER has extensive reserves of mineral resources, the extent of which has not been fully explored. The most important mineral, for general industrial purposes, is iron; although adequate enough to sustain an integrated iron and steel complex it is still not exploited for regional consumers. This is important as it is badly needed for comprehensive, regional, development purposes. This was an argument that was promoted during the Soviet period, and it appears now and again in the present debate. What has been exploited, in spite of distance and the severity of the climate, are diamonds. The FER has, until quite recently, functioned as a monopoly supplier of diamonds in the Soviet Union and later in the Russian Federation. Gold, boron and tin are exploited, due, on the one hand, to their high value, and on the other, to their relative scarcity in other parts of the former Soviet Union. Although still of economic importance, it can be argued that the openness of the Russian economy makes a strategy of self-sufficiency no longer an issue of importance. Considerable zinc deposits are to be found in the area, reported to be the richest in the entire Russian Federation. Other important, but mostly undeveloped, mineral resources include mercury, phosphate, apatite and graphite. FER is today not

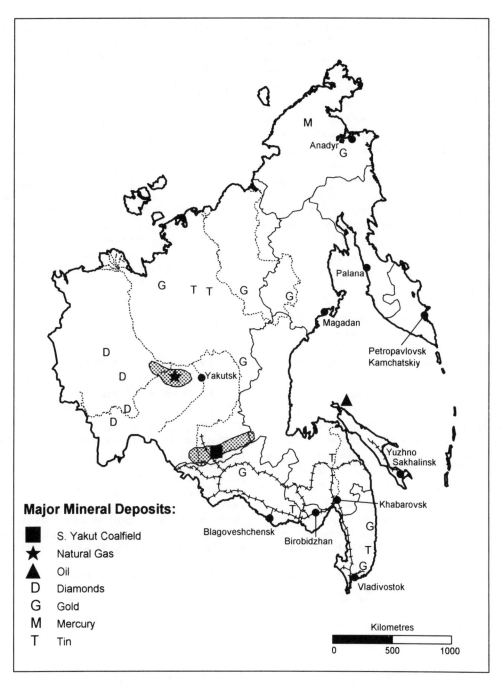

Map 12.4   *The resource base of the Russian Far East*

a manganese-producing region although some reserves are to be found within its territory. Aluminium ore is to be found. For copper the FER region is today of only marginal productive interest, although the extended territory includes the Udokan copper reserves, where a Russian-led international conglomerate has recently been given exploitation rights. The potential is

197

substantial. The region also provides tungsten, molybdenum, mercury, bismuth, fluorspar and lead to the national economy. It is not clear whether the production of these minerals is competitive on the international market, and the picture will remain blurred for some time to come. Building materials such as limestone for cement, clay for bricks, tile and gravel for roads and concrete are also found. The latter are of regional or local interest only. The mining share of the FER output volume is 30% compared with the Russian Federation average of 11% (see note 1).

The fishing industry is today, and will probably remain, one of the most important branches of economic specialisation in the FER. It supplies the Russian Federation with about 60% of its total catch. Biological resources in the 320 km (200 mile) sea zone are estimated at 25.8 million tonnes including 23 million tonnes of fish (*Investor's Guide, Russia 92*, 1993, p. 59). Another sector of central economic importance is forestry where the FER provides roughly 10% of total Russian production including a large share of valuable timber. FER is also distinguished by a high proportion (Russian Federation) of non-ferrous metallurgical industries (between 14 and 100% depending on the mineral) (Barr, 1990, p. 114).

The FER's share of total industrial output of the Russian Federation broken down into sectors is shown in Table 12.2. This table does not cover one very important part of the picture and an analytical variable with clear implications for the FER's role in the national economy, and therefore also with a political dimension. That is the degree of monopoly in production which creates a degree of dependency of regional production for the Russian Federation economy. Vice versa, there is the dependency of the FER on interregional imports. More up-to-date statistics are hard to find, but an outline is given in the excellent work by Schiffer (1989, pp. 202–4):

... the Far East produces 20 per cent of the all-Union output of foundry machinery, almost as high a percentage of the all-Union output of bridge-girder electric cranes, 7 per cent of the all-Union output of cranes for the automobile industry, 5.2 per cent of the all-Union output of power transformers, and more than 4 per cent of the all-Union output of forging and pressing machinery. Owing to the all-Union orientation of much of its output 75–98 per cent of MBMW [machine building and metal working] production of automatic machine tools, power transformers, hoisting and transporting equipment, energy generating machinery, and fishing and trade equipment are exported from Far Eastern MBMW plants to other regions of the USSR.

The FER shipped out a total of about one-third of the value added in the region, and, of this, 93% went to the internal Soviet market (Thornton, 1995, p. 82). A very high dependence on imports from other regions in other sub-branches of the machine-building industry was also characteristic. For example:

... all tractors, cars and trucks are imported, foundry equipment – 70 per cent, cranes – 56 per cent, lathes – 80 per cent, and diesel and diesel generators – 50 per cent. Eighty per cent of its needs of machine tools come from other regions with half from European Russia alone (Rodgers, 1990, p. 235).

Table 12.2   *The Far Eastern economic region's share of the Russian Federation's industrial output (see note 1)*

Sector	Share of output (%)
Electrical power engineering	5.0
Fuel industry	3.4
Metallurgy	6.9
Machine-building	2.8
Wood-chemical industry	3.9
Construction materials industry	8.8
Light industry	1.6
Processing industries (food industry including fishing)	11.0
Total	5.4

Source: *Investor's Guide* (1993, p. 59).

All in all, shipments into the region equalled about 60% of regional value added, which included a large share of military hardware. Agriculture is limited by climatological factors but is also underdeveloped; 202.9 million ha is under the plough and yields only 7% of the gross product, which means that it provides for no more than 40% of the amount of the agricultural consumer goods that are actually consumed in the region – 95% of all non-food consumer goods were also imported and 80% of all these imports emanated from the former Soviet Union (Thornton, 1995, p. 82).

Due to great differences in climatic conditions and historical and economic factors, the structures of industrial production vary substantially in FER subregions (see Table 12.3). The regionally differentiated economic profiles touch upon the theme of short- and long-term interests, as they relate to the centre–peripheral and intra-regional political processes. It also touches upon the central questions relating to the attractivity of the Pacific Rim. Before we enter this field it is necessary to discuss some of the institutional and structural obstacles that stand in the way of any kind of economic development in the territory. As has been noted, the FER could be described as a territory saturated with riches. The potential stands, in general, in stark contrast with chronic lags in the development and exploitation of these resources.

The main argument raised against a comprehensive SIB (Siberia)/FER development is the structural and geographical obstacles that appear and primarily the impact they have on the cost of production, transport and consumption. It is true that the specific SIB/FER context, to a varying degree, creates special extra costs. Promoting this

Table 12.3 *The structure of industrial output in administrative territories of the Russian Far East in 1991*

Sectors	Far East %	Sakha (Yakutia) %	Primorskiy kray %	Khabarovsk kray %	Amur oblast %	Kamchatka oblast %	Magadan oblast %	Sakha oblast %
Electro-energy	3.5	3.2	2.5	3.1	6.1	4.8	6.0	2.2
Fuel	3.8	6.1	1.0	6.6	3.3	0.0	1.2	5.4
Ferrous metal	1.1	0.0	0.1	4.1	0.1	0.0	0.1	—
Non-ferrous metal	19.4	63.2	4.0	5.0	13.5	0.6	59.4	0.2
Chemicals and petro-chemicals	1.8	0.0	3.0	4.0	0.1	0.0	—	0.4
Machine-building and metalworking	15.4	2.2	21.0	28.7	13.6	8.5	6.1	4.6
Forest and paper work	9.2	3.7	6.4	11.7	10.5	3.1	2.0	28.8
Building materials	6.5	7.2	7.7	6.2	7.2	5.6	4.8	5.4
Glass and ceramics	0.3	—	0.4	0.1	1.1	—	0.4	—
Light industry	4.2	1.3	4.6	7.5	5.1	2.0	2.6	1.8
Food processing	31.8	12.5	45.2	18.9	33.3	73.3	16.4	47.7
Flour processing and cereals	1.9	0.0	2.5	2.5	5.4	0.6	0.0	2.7
Other branches of industry	1.2	0.6	1.6	1.6	0.7	1.5	1.0	1.5

Source: Mikheeva (1993, p. 16).

fact are, among other things, the sparse infrastructure, the necessity of adapting technology to the harsh environmental circumstances and, most important, the need to increase and stabilise the labour force. The picture presented, in a multitude of pre-*perestroika* sources, of raw materials and energy available at low cost (*in situ*) in SIB/FER is valid. It is, however, of far more interest to discuss whether these, supposedly, low costs can be carried through the production process in all its stages. What is indicated here is that the general lack of development is in itself the fundamental cost-raising dimension. Some of these factors, like manpower, infrastructure and construction, will be discussed here. Excluding technological, quality and environmental aspects makes the total picture incomplete and could only be blamed on the lack of space.

Manpower

It is arguable that the most serious obstacles to Far Eastern development are the lack of a sizeable labour force and the proven inability of the Far East's rugged living standards and lack of amenities to inhibit the permanent settlement of migrants from other parts of the country (Helgeson, 1990, p. 58).

In its concentrated form this quote does give an outline to the problem of a general manpower deficiency, determined primarily by the low absolute figures of the able-bodied male population. The demand for manpower has been and is still high for reasons relating to the low level of automation and the industrial profile where a preference for male workers has dominated the picture. A flow of workers from the rural areas has not been the solution, as the manpower problem in the countryside is even worse. Numerous complaints about the extreme lack of qualified workers, especially with expertise in severe conditions, have been heard. Still, the picture is, as indicated, only halfway true as unutilised labour reserves in the FER were 1.5 times greater than the Soviet average. Low female participation in the labour force explains this anomaly. During the Soviet period, efforts to recruit manpower to the region met with remarkably little success and even so, problems of composition and high turnover (long adaptation time, low education and skill levels) led to low productivity. Today, recruitment campaigns, based on special wage levels and other extras, are something of the past and have been replaced with market-determined wages. Interindustrial manpower flows are, however, not taking place on a major scale and a general outflow from the region is part of the picture. In 1993 the outflow exceeded the natural population growth by 24 000, with 6.9 times as many people leaving the area as in the preceding year (Russia's interest . . ., 1995, p. 22). A positive contribution that is slowly developing and which does have more far-reaching consequences is labour reserves in regions considerably closer and more environmentally similar to the FER than the western regions of the Russian Federation. With political normalisation developing with its neighbours "joint activities involving labour from Korea, China and Vietnam are being explored at a dizzying pace" (Helgeson, 1990, p. 75). Foreign workers have been present in the region since the mid-1960s. Today, Vietnamese and Korean workers can be found in timber extraction, Chinese and Korean labourers in agriculture and in the construction sector and a lot of activities involving foreign manpower in the service and small-scale commercial sectors. In spite of all the action taking place, a solution on any larger scale does not seem feasible. There are probably around 100 000 Chinese living illegally in the FER, while opponents of "Chinese colonisation" talk of 2 million illegal immigrants (*Transition*, 8.9.95, pp. 5–6). Because of growing social tensions, public opinion is running against the large-scale recruitment of unskilled workers from Asian countries. Pavel Grachev still complains that "Chinese citizens are peacefully conquering Russia's far east" (*Transition*, 8.9.95, pp. 5–6). A case in point is the special administrative regime introduced in the Primorskiy *kray*, which decided on special treatment for Chinese visitors (Kirkow, 1995, pp. 924–5). Restrictions to certain sectors of the economy and/or subregions have been suggested as a possible solution to this problem.

Infrastructure

Transport in the FER plays a more crucial role for the economy than in any other region of the country. More investment in transport is needed to achieve a given economic objective than elsewhere. Transport costs constitute a higher share of total delivered cost for most products. The burden of extra costs for regional production and consumption is a crucial development factor: ". . . the share of loss-making enterprises in the Far East (28.5 per cent in 1992) is far greater than elsewhere in Russia, which must largely be due to considerable freight costs" (Russia's interest . . ., 1995, p. 21). Transport workers constitute a high proportion of the total workforce and the number of transhipments per tonne of goods delivered was at least double the union average (North, 1990, pp. 185–8). The reason for this is quite obvious with the distribution and sparseness of population and economic activity and the distance from European core regions of Russia. In order to maintain a minimum of economic activity and consumption, transport distances, by necessity, are long and the freight volumes handled relatively small. The FER has the sparsest transport network in the country, by any measure relating route length to area (North, 1990, p. 191). This means, in more general terms, that the FER is very vulnerable to increases in transport costs. So, when a tonne of cargo shipped from Japan to the FER is three to five times less than the expense of bringing the cargo from the European part of Russia, the ultimate outcome is obvious (Russia's interest . . ., 1995, p. 23).

Other infrastructural elements, housing and other social and cultural facilities have had and still have an important impact on the manpower situation. The situation has, in local interviews, been described as extreme and the need for resource allocation in these sectors is very high. A potential foreign investor is also likely to scrutinise the situation as regards reliable water provision, sewers, modern communications, premises, hotels, exhibition grounds, advertising and information services. To put it mildly, there is a long way to go in order to reach an acceptable level of provision.

There is another infrastructural element of importance and that is the institutional structure. Regional performance in the transformation from a centrally administered to a market or semi-market economy varies substantially. Compared with the rest of the RF, the SIB/FER's economic activities are still, to a higher degree,

controlled by politicians and bureaucrats. (See the informative article by Kirkow, 1995.) The lag in foreign interest and resultant foreign investment can partly be explained by this. Lately there have been signs of increased activity in this field, which also shows in an increased investment activity.

Construction

The construction sector in the FER is characterised by a number of deficiencies that have had and still have a severe impact on economic development. Due to the lack and uneven development of the construction base and its irrational distribution across the region, implementation of plans for large-scale projects have sometimes been postponed more or less indefinitely, favouring the relocation of these projects to other more "suitable" sites in other regions. During the Soviet period central ministries did not see any reason for any strategic thinking in the construction sector. Irrational distribution of the construction base linked with the important distance factor, resulted in large volumes of uneconomic haulage. Equipment is more outdated that the average for the Russian Federation. The status of the construction base necessitates large and costly import volumes in spite of the existence of local but unexploited resources. Physical harshness also has a substantial impact on productivity, as normal all-year construction work is seasonal and outdoor work is further limited in large parts of the FER.

Potentials and interests

It is becoming more and more evident that an important dimension of future development lies in the sphere of struggle for economic and political influence and control. The main conflict is found in centre–peripheral relations, although intraperipheral and regional/local dimensions are also evident. A general ambition towards economic reform, and hereby economic growth, is the basic objective in the background to the present development. Since the early days of *perestroika*, a major element in this has been a general, but unclearly specified, decentralisation. The substance, from the viewpoint of Moscow, has focused on gaining efficiency and growth within the economic sphere, by decentralising decision-making. The new openness in public debate and in the media and the efforts to promote democratic processes and procedures gave a new framework for political activity. This activity, in its first instance, took nationalistic and ethnic features and led to the dissolution of the Soviet Union and a process of disintegration within the Russian Federation. The breakup of the union has been manifest for a number of years now, although a pendulum effect seems to be taking place between Moscow and some of the former union republics. During the last couple of years the tug-of-war between Moscow and the peripheral regions has intensified. SIB/FER regions have had a prominent role in this process.

Before we venture into this discussion, it is important to note that the SIB/FER definition, although relevant, is in many cases too general, and subregions (SIB/FER administrative and other entities) appear with different tendencies and roles. This phenomena has not come out of the blue as even during the pre-*perestoika* period the centralism of state authority was complemented by the strong positions of local élites, who ran the regions as their own respective fiefdoms. The influence of local leaders has now increased, when they confirmed their legitimacy and independence in popular elections. A special SIB/FER dimension to this regionalism is that it is not primarily ethnic, national or confessional, but economic (de Souza, 1994). The functions of the present-day centre are not clearly manifest and there is no real executive structure to implement its decisions, giving regional authorities the illusion that the central government is "redundant". One of the possible arguments for the centre's activity in Chechnya was and is to increase its authority. It is not, however, the relative weakness of direct political control from the centre that is relevant. Short-term economic priorities and necessities create a situation where traditional resource extraction combines with an orientation towards the European parts of the Russian Federation. As has been mentioned, the Far Eastern economy has a quite high dependence upon the rest of the economy. Moscow's ability to deliver crucial resources seems to diminish from day to day. With the old supply system breaking down, and new suppliers found in other geographical directions, the periphery's interest in the centre's future ability to provide resources weakens:

> Given the present conditions, stimulating development in the Far East will now take either a massive influx of centralised investment (an unlikely thing while the state budget continues to show a deficit) or more active integration into the APR and NEA economies (Russia's interest . . ., 1995, p. 6).

Centre-based priorities can also stand in stark contrast to local/regional ambitions. Just to take one example from hundreds of similar issues: the Khabarovsk *kray* authorities are planning gold and silver mining operations on the banks of the Okhotsk Sea, while the centre's administration prefers to concentrate on promoting the tourism and leisure industry that caters to foreign-currency-paying vacationers. There is a feeling of colonial exploitation that is quite widespread. This stands in conflict with long-term ambitions of the SIB/FER regions towards a more comprehensive, diversified and self-sufficient strategy. In the FER context, the potential riches and pull factors of the Pacific Rim also stand out as prominent. Further enhancing aspects of regional self-sufficiency are the geographical dimensions of economic factors. A good example is the dramatic increase in domestic transport costs (in 1992 the tariffs were on average increased 30 times (Manezhev, 1993, p. 5)) making the internal (Russian/European) market lose some of its "pull" and relocating both imports and exports to territories with a close proximity. Distance, combined with transport cost factors, results in centrifugal processes, especially when it comes to the development of wholesale pricing.

Locally and regionally, new kinds of coalitions are being created to fight the new "central powers" and also to exploit the relative weakness of the centre. The primary question has all the time been the increase of economic autonomy, even though voices have been heard demanding the establishment and creation of a sovereign Far Eastern republic or a Siberian republic. There is a clear manifestation of frustration regarding the slowness of the decentralisation process. This has raised the political temperature substantially, which means that declarations of sovereignty or autonomy win supporters. Regional organisational activity with examples as the "Siberian Agreement" formed in 1990 and uniting 19 administrative entities in west and east Siberia and the establishment of the Far Eastern Association for Economic Co-operation are symptomatic. Political parties have also been formed with sovereignty and/or autonomy on their agendas like the Far Eastern Republican Party of Freedom and the Party of Revival.

In the conflict between the Yeltsin administration and parliament the regions played the two contenders against each other and through this won privileges and subsidies. The referendum on a new constitution and the election to a new parliament was to become a focal point for the reinforcement of Moscow's authority, in view of extended presidential power. The outcome of the election to the parliament, however, did not solve the issue of contention. In spite of the personal endorsement of Boris Yeltsin and presidential power, the role of the person is substantially reduced, but the position is in itself strengthened and exploited by the President's entourage. The government, in turn, with Viktor Chernomyrdin in the seat of Prime Minister, has increased its authority. At present, issues that could be interpreted in terms of regionalism, separatism, autonomy, etc. are temporarily taking a back seat. The importance of the issues at hand in the existing core–periphery relations will, however, bring them forward again.

As a base for the federation, regionalist tendencies could become a serious threat to the future stability of Russia. Emphasising, as is done here, the disintegrative forces in the Russian Federation and in SIB/FER relations with Moscow, means a number of caveats have to be acknowledged. Most prominent of these are old economic structural relations which could not, in the short term, be replaced by new linkages, by potential budgetary allocations and by other institutional linkages, like the Russian military sector.

As has been mentioned, to analyse SIB/FER development on an aggregated level does not contribute much to the analysis. It focuses too much on generalised strategic considerations and does not recognise the diversity of regionally/

locally based strategies. Aspects of export capacity, of contribution to national (primarily European Russian) industrial development and degrees of local or regional diversification, have to be evaluated in order to reach an understanding of the reasons behind these strategies, their political implications and their possible result in the future. In many cases, the process of independence/autonomy rests on the assumption that the region will be better off if it can take control over its own riches and direct the income flows from those riches directly to the needs of the region. With the weakening of the centre's ability to provide the basic aspects of supplies, there is an increase in local or regional need for control over sources generating vital hard currency. The patterns of single-industry dominated cities and urban settlements increases the importance of the impact on specific branch development on the local, settlement, level as well. This argument leads, indirectly, to an argument that declared ambitions for independence or sovereignty, which dominates the overall picture, are a means to extend this control. The real ambition, however, is limited to an extended regional/local control in economic terms. The important distinction that should be made is that within the realms of the general SIB/FER development, there exist a number of subregional and subsectoral strategies and tendencies, based on the specified economic (structural and potential) profiles. Variables of importance here are: raw material production and reserves, defence industry and necessity of conversion, level and necessities of the consumer-oriented sector, capacities and ambitions for the processing and manufacturing sectors (the latter broken down for specific subbranches). This should be further disaggregated down to specific economic and administrative entities like *oblasts*, rayons and in specific large-scale projects.

It is possible to identify a potential polarisation between regions/localities with resources of one kind or another which are vital to the Russian Federation and/or the international market and others with only limited resources. There is now a sharp competition between administrative entities for infrastructural projects such as international airports, for the status of free economic zones and for the establishment of consular facilities, as well as for more symbolic phenomena, like having the first four- or five-star hotel. It is reported that

Today, it (Khabarovsk) is fighting to retain its status as the regional hub against competition from Vladivostok and the port's neighbour, the free-enterprise zone of Nakhodka (*Far Eastern Economic Review*, 8.7.93).

Basically, the competition is about the potential flow of future investment resources and trade. In this conflict everything seems to be permitted. As has already been mentioned, the central administration has been playing this card in order to gain the upper hand, and continues to do so. Environmental and health hazards are usually not analysed as factors stimulating integrative or disintegrative tendencies. The severity of the problems in the SIB/FER context and the link between Moscow's policies in the past and the present regarding the environmental situation promotes, at least, two disintegratory phenomena. The first one is the impact (one could even say environmental disaster) in areas inhabited by the northern ethnic minorities in the form of health effects and disruption of the traditional economic activities (Feshbach and Friendly, 1992). Although these groups and peoples do not have any political clout worth mentioning, the areas inhabited by them are included in the future resource base, and the treatment of these peoples can become a political nuisance for the Russian Federation, both in a national and international context. The South Korean Hyundai's Svetlaya project (timber extraction) is a case in point, although it was also used to improve the power-base of the governor.

With reduced international tension and less aggressive foreign policies, the importance of maintaining the military capability built up during the cold war has diminished. Apart from the international arena, the military has still the potential of playing an important role in the internal development of the Russian Federation and the CIS. During the Gorbachev years, it used to be emphasised that the military sector was the only remaining power that could, by its organisation and territorial coverage, keep the union together. Later, during the early Yeltsin "era" the same was said about Russia and the Russian Federa-

tion. Nowadays, the message is more blurred. The question is where a further weakening of Moscow's authority could lead. If disintegration gains momentum aiming for political independence, regionally based armed forces in contention with the Russian Ministry of Defence could become a reality. Comments have been heard that plans to strengthen regional commands could undermine central Russian military control. It has even been mentioned that the present and future military territorial organisation, replacing the present one, is a rational way of solving the conflict of ethnic versus economic delimitation (Allison, 1993; *Moscow News*, 1993, p. 22).

Economic and geographical processes and patterns

Returning to the title of this chapter, emphasising the concept of Russia's "gateway" to the Pacific, on the basis of what has been discussed so far, seems somewhat misconceived. It also seems evident that whatever metaphor we use, it will not, to a full extent, cover the multifarious position of the SIB/FER. Geostrategic location, economic potential and structural impediments, infrastructural position, impact on internal power politics, etc., all the variables seem to add up to a list of potentials or barriers to development, where the coming decades will point out which processes and patterns that were and will be decisive in forming its comprehensive future profile. The situation should not, however, prevent us from launching a tentative analysis. The analysis should, however, be regarded as an ongoing process, developing through time and not in any way finalised.

Concluding some of the more important dimensions being used in this chapter, the nature of the centre–periphery power struggle, linked to the spontaneous and regulating forces of the present economic development, also linked to the appearance of outer forces exerting a powerful attraction, are the fundamental processes and patterns which are already appearing in the FER today. Of fundamental importance is, naturally, the political and macroeconomic framework, in a wider global context.

The metaphor of the "gateway" is clearly evident in policy formulation, in a manifest ambition by the centre to turn the Asiatic parts of Russia eastwards, to benefit from the development processes of the Pacific Rim. The "gateway" is also an entrance, an invitation to take part in the reconstruction and economic future of the whole of Russia. The Pacific community, or relevant subdivisions of this entity, will, by reason of Russia's returning international importance, successively play a more prominent role in Russian foreign policy considerations which may be expected to affect every major development in the FER region in the foreseeable future, i.e. the FER is not only an economic option for the Russian Federation, but also increasingly a card to be considered in foreign policy considerations. As has been mentioned, this mixture of general political ambitions, of Pacific presence, and an export-base strategy, has a problematic reflection in internal conflicts regarding who should control the resources exploited and incomes generated. The policy or strategy itself does not stand undisputed as arguments are raised to give priority to the national needs of raw material and energy resources. Finally, regional and local demand for resource and income control are, in the best of cases, promoted by the idea of local and regional development and diversification. To assume that the future will see a mixture of these policies and strategies is easy to do and most realistic. It is, however, too early to state where an emphasis will appear. The basic fact is, although not yet a tendency, that the FER is increasing its trade with the Pacific at a more intensive rate than the increase in Russian foreign trade in general. In 1993 the figures were 18.4% for the FER compared with 1.4% increase for Russia. It has also to be noted that the increase is for raw materials, which indicates a continued and developed role of a raw material appendage to the Pacific economic community. This role also seems to be the realistic one in relation to Russia and the former Soviet territory (Russia's interest . . ., 1995, p. 21).

Leaving the policy level and entering into the reality of resource flows of different types, there is a clearly discernible tendency of severing the links with the centre. Whether this is a conscious, economically based decision or just a spontaneous breakdown of links is not important in itself. There is an increased intraregional (FER) interaction which is explained by economic forces like increasing transport costs, but also by a number of non-economic factors. Starting out from 1992 the reorientation of the FER economy towards the Pacific began. This development is severely hindered by a number of factors. From those already mentioned in this chapter, the centre–periphery dimension clearly stands out as paramount. The balance-of-payments problem, significant for the Russian Federation, gives central priority towards building up an export orientation, with the primary purpose of obtaining foreign currency on a regional level, as a necessity. The exports from the FER are exclusively based on strategic raw materials under surveillance from the central government. Control from the centre seems, however, to be more and more relaxed.

Interests in the Pacific Rim towards the FER have their special priorities and nationally induced priorities for exports of the processing industry clash with local/regional interests, as they bring with them a "burden" of indirect economic phenomena which, unhindered, will create severe structural problems of closures of enterprises and large-scale unemployment. That some kind of protectionism will appear is logical, when the problems begin to affect the social and political situation in a more direct way. What is unclear here is whether centre and periphery will work together on these issues. Otherwise the centre–peripheral contention will find a new outlet.

Everyone in the FER administrative entities, especially in the centres of Khabarovsk, Vladivostok and Sakhalin, is competing with one another rather than developing an *esprit de corps* as a united Far Eastern region. The success of one or the other or, more important maybe, the uneven distribution of economic opportunities and resources, will form the basis for future economic structures, and thereby regional imbalances. There is a clear geographical context for the potential, externally generated, pull-force. We can distinguish a pattern, within an economic framework, in, for example, the following categories: geographical situation (maritime or landlocked, south or north orientation); transport relations; as well as resource profile and complementarity in relation to each subregion's production profile with neighbouring countries.

A geographical pattern that is also discernible today is that of a growth pole orientation. Major cities and settlements in the existing system will be the focus of foreign interest, creating a resource exploitation hierarchy. This will be complemented by a dimension of penetration primarily from the ocean and from the Trans-Siberian Landbridge. It is in the northern areas, where considerable natural resources have been and are to be discovered, that an important part of the future economic development of the Far East will take place. The southern zone will become the base camp for this development. Based on fragmentary information this process can already be distinguished. Tomorrow will see the formulation of development strategies along these lines linked to some dimensions of short- and long-term "planning". Regional (and eventually national) strategies will be directed towards the dissolution of the most important bottlenecks, both in infrastructural and production terms. Investments into an increased metallurgical capacity in the FER are a pertinent example. Business infrastructure, such as accommodation and telecommunication, is an immediate and acute necessity, where a wide range of activities is already taking place. Intensive development in border trade is already promoting a range of necessary changes and development of auxiliary activities. Another logical dimension of interest is the increased attention towards the transport network, where increased trans-Russian product flows will demand substantial improvements in port-handling capacities,[4] major transport arteries and feederline networks. The share of the Pacific Rim countries in the total trade with the Russian Federation increased between 1991 and 1992 from 12 to 20%. However, trade generated in the FER constitutes not more than 20% of this (Mikheeva, 1993, pp. 13–14).

Parallel to the need to disaggregate regional analysis stands the need to distinguish a Pacific impact in multilateral terms, but even more so in bilateral relations. Prominent actors such as Japan, the USA, the People's Republics of China, Korea and Vietnam, the Republics of Korea and China, Hong Kong and Singapore all show an increasing economic interest in Russia in general and an ambivalent but increasing interest in the FER.

The international environment is also the dimension where individual economic actors, enterprises, organisations and so on, formulate their individual strategies towards the FER. Long- and short-term profitability is the keyword. Its prerequisites are found, apart from in the general political climate, in the specificities of the branch-related economic situation, nationally determined financial conditions, FER, Russian and Soviet-related production and economic conditions. Its outcome, by the nature of its orientation, is of primary importance on a local or subregional level, on specific branch or subbranch development. Its impact is of more far-reaching importance when contributing to diversification and/or extension of the economic structure of the FER. At this date it is unanalysable. Development strategies and planning will eventually include preferential treatment of sectors that can show competitive advantages of evident and potential complementarity with the neighbouring nations. The already appearing profile as a shipment-transit territory will have an impact on these priorities and especially resource allocations from the centre. A special role and dimension also are the introduction of preferential treatment for joint ventures in the FER and the establishment of "free" or "special" economic zones, although this path of development is still bogged down by bureaucratic obstruction.

This development, of course, will require a combination of huge short-term infrastructural investments and, likewise, huge long-term capital investment of a more generalised nature. As investment resources will, in the face of demand, be extremely scarce, this will be the major bottleneck for the future, whatever route this future takes. The question is not "if" but "how", "when", "where" and "by whom" the development of the FER should and will continue. Questions regarding the "how" have to turn to one point of central importance for economic development in its FER context, and that is, what could be termed "thresholds". The economic rationality of project evaluations and investment calculations based on existing economic and other structural phenomena does in many cases lead to the conclusion of unfeasibility. Today the concerns, as for foreign participation, are primarily for the political and institutional prerequisites. With the negative aspects of this out of the way or substantially reduced, attention will turn to general infrastructural phenomena. The idea of a critical economic mass should be a recurrent idea in the overall analysis.

Notes

1. The region has seen a reduction in population during the last couple of years, which will be commented upon in the continuing text.
2. All in all, 1995 has shown a stabilisation of the inflation on a manageable level.
3. The source (*Investor's Guide 92*) does not indicate which year is referred to. A reasonable assumption is, however, 1991.
4. This is duly noted by the EBRD which contributes to the development of the Nakhodka harbour.

References

Allison, R. (1993), "Russian defence planning: military doctrine and force studies", in *Lectures and Contributions to East European Studies and FOA*, Stockholm: The Swedish National Research Est.

Barr, B.M. (1990), "Forest and fishing industries", in Rodgers, A. (ed.), *The Soviet Far East – Geographical Perspectives on Development*, New York: Routledge, pp. 114–62.

Bradshaw, M.J. (1988), "Soviet Asian–Pacific trade and the regional development of the Soviet Far East", *Soviet Geography*, 29(4): 367–93.

Dalniy Vostok Rossii – Ekonomicheskoye Obozrenie (1993), Moscow: Progress-Kompleks.

De Souza, P. (1994), "Disintegration of Russia – integration into the Pacific?", in Edström, B. (ed.), *Current Developments in Asia Pacific*. Stockholm University: Centre for Pacific Asia Studies, pp. 4–48.

Dienes, L. (1988), "A comment on the new development program for the Far Eastern Economic Region", *Soviet Geography*, 29(4): 420–2.

Feshbach, M. and Friendly, A. Jr (1992), *Ecocide in the USSR*. New York: Harper Collins.

Helgeson, A (1990), "Population and labour force", in Rodgers, A. (ed.), *The Soviet Far East –*

Geographical Perspectives on Development, New York: Routledge, pp. 58–82.

Holzman, F.D. (1957), "Soviet Urals Kuznetsk combine: a study in investment criteria and industrialisation policies", *Quarterly Journal of Economics*, 71(3): 368–405.

Investor's Guide, Russia 92 (1993), Moscow.

Kirkow, P. (1995), "Regional warlordism in Russia: the case of Primorskii krai", *Europe–Asia Studies*, 47(6): 923–47.

Kovrigin, E.B. (1986), "The Soviet Far East", in Stephan, J. and Chichkanov, V.P. (eds), *Soviet–American Horizons on the Pacific*, Honolulu: Univ. of Hawaii Press, pp. 1–16.

Manezhev, S. (1993), *The Russian Far East*. Post-Soviet Business Forum. London: Royal Institute of International Affairs.

Mikheeva, N.N. (1993), "Structural changes and integration of the Russian Far East into the Pacific market", paper presented at the 33rd European Conference of the RSA, 23–27 August.

Mote, V. (1983), "Environmental constraints to the economic development of Siberia", in Jensen, R.G., Shabad, T. and Wright, A.W. (eds), *Soviet Natural Resources in the World Economy*, Chicago: Chicago University Press, pp. 15–71.

North, R.N. (1990), "The Far Eastern transport system", in Rodgers, A. (ed.) (1990), *The Soviet Far East – geographical perspectives on development*, pp. 185–224.

Rodgers, A. (ed.) (1990), *The Soviet Far East – geographical perspectives on development*. New York: Routledge.

Russia's Conquest of Siberia. 1558–1700. A Documentary Record. (1985), Edited and translated by Dmytryshyn, B., Crownhart-Vaughan, E.A.P. and Vaughan, T. Portland: Western Imprints, The Press of the Oregon Historical Society.

"Russia's interests in Northeast Asia and the prospects for multilateral cooperation with NEA countries in promoting the development of Russia's Far East (1995), *Far Eastern Affairs*, No. 3: 4–46.

Schiffer, J.R. (1989), *Soviet Regional Economic Policy – the East–West Debate over Pacific Siberian Development*. London: Macmillan.

Starr, S.F. (ed.) (1987), *Russia's American Colony*. Durham: Duke University Press.

Thornton, J. (1995), "Recent trends in Russia's Far East", *Comparative Economic Studies*, 37(1), 79–86.

Weiss, K.G. (1989), "The strategic triangle and Sino-Soviet crises", doctoral dissertation, the Johns Hopkins University.

13
Transport in a new reality

Robert N. North
University of British Columbia, Canada

Introduction

Political disintegration and the collapse of the command economy have created a new environment for transport in the former Soviet Union. This chapter describes that environment and the responses of the transport industry to it.

Environmental changes and responses cannot be understood properly without knowing what went before. Analyses of transport in late Soviet times can be found in North (1991, 1995). Here it must suffice to point out that the transport system entered the post-Soviet era in poor condition. A lack of investment during the 1970s and 1980s had left it with obsolescent and run-down equipment. A late Soviet-era attempt to transfer financial responsibilities from the controlling ministries to such low-level organisations as ports and regional railways had exposed the inadequacies of the tariff system and left many operators with financial worries and declining morale. The structure of transport suggested future problems in that its most poorly developed component, road transport, was a major facilitator of economic growth in Western economies, and in that it lacked intermodal operators. Indeed, the various modes were much more inclined to compete than co-operate, despite official trumpeting of a "unified transport system". And a decline in traffic after 1988, following years of continual growth (Tables 13.1 and 13.2 and Figure 13.1) was both a blessing and a curse. If traffic had continued to grow, a deteriorating system could probably not have carried it all. But declining traffic meant declining revenues, with which many operators could not cope.

Trends in traffic

Transport statistics from different sources for the post-Soviet period vary somewhat, reflecting practical problems of economic and bureaucratic transition, but the general picture is reasonably clear. The decline in traffic, mentioned above, accelerated after the dissolution of the USSR. It slowed down only in 1994 and flattened out in 1995 (Table 13.3). Decline was not evenly spread,

Table 13.1 *Traffic by transport mode, USSR, 1913–90 (common carrier, % of tonne-kilometres)*

	1913	1940	1950	1960	1970	1980	1990
Railways	60.6	85.1	84.4	79.8	65.2	53.1	46.9
Maritime shipping	16.1	4.9	5.6	7.0	17.1	13.1	11.9
Inland waterways	22.9	7.4	6.5	5.3	4.5	3.8	2.9
Pipelines	0.2	0.8	0.7	2.7	7.4	28.0	36.5
Motor transport	0.1	1.8	2.8	5.2	5.7	2.0	1.7

Sources: *Narodnoye khozyaystvo SSSR v 1960 godu*. Moscow: Gosstatizdat, 1961, p. 531; *Narodnoye khozyaystvo SSSR v 1980 godu*. Moscow: Finansy i statistika, 1981, p. 293; *Narodnoye khozyaystvo SSSR v 1990 godu*. Moscow: Finansy i statistika, 1991, p. 582.

Table 13.2 *Traffic reduction, USSR, 1988–90 (common carrier, % decline in tonnes dispatched)*

Transport mode	% Decline	Transport mode	% Decline
Railways	5.9	Motor transport	8.3
Maritime shipping	10.9	Air transport	12.1
Inland waterways	3.2	All transport	6.7
Pipelines	1.3		

Source: *Narodnoye khozyaystvo SSSR v 1990 godu*. Moscow: Finansy i statistika, 1991, pp. 576 and 582.

Geography and Transition in the Post-Soviet Republics. Edited by M.J. Bradshaw.
© 1997 M.J. Bradshaw & Contributors. Published 1997 by John Wiley & Sons Ltd.

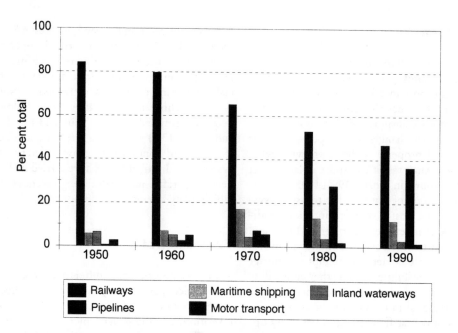

Figure 13.1 *Traffic by transport mode, 1950–90, common carrier (% of tonne-kilometres) (for sources see Table 13.1)*

Table 13.3 *Freight dispatched, Russia, 1990–94[1] (million tonnes)*

	1990	1991	1992	1993	1994
Common carrier					
Railways	2 140.0	1 957.0	1 640.0	1 348.0	1 058.0
Maritime shipping including cabotage	112.0	104.0	91.0 16.9	83.0 11.9	69.0 7.9
Inland waterways	562.0	514.0	308.0	215.0	155.0
Pipelines	1 101.0	1 042.0	947.0	873.0	800.7
Motor transport	2 941.0	2 731.0	1 862.0	1 109.0	652.8
Air transport including regular carriers	2.5	2.2	1.4 0.7	0.9 0.5	0.7 0.4
All transport	6 858.5	6 349.2	4 908.3	3 625.7	2 959.4
Non-common carrier					
Industrial railways	6 357.0	5 789.0	5 101.0	4 050.4	3 507.2
Motor transport	12 406.0	11 957.0	10 597.0	9 091.0	6 947.2
All carriers					
Total	25 621.5	24 095.2	20 606.3	16 767.1	13 413.8

Notes: 1. Many of the figures given vary slightly among sources.

Sources: *Rossiyskaya Federatsiya v 1992 godu. Statisticheskiy yezhegodnik*. Moscow: Goskomstat Rossii, 1993, pp. 538–44; *Sotsial'no-ekonomicheskoye polozheniye Rossii*. Moscow: Goskomstat Rossii, 1 *Rossiya v tsyfrakh*. Moscow: Goskomstat Rossii, 1995, p. 267; Gol'ts, G. and Filina, V. "Strukturnyye sdvigi na transporte", *Avtomobil'nyy transport*, 1995, No. 5: 22, 23; Bagrov, L. "Rechnoy transport: itogi, zadachi", *Rechnoy transport*, 1994, No. 2; 3.

indeed it was accompanied by major shifts away from the Soviet pattern of spatial linkages. We can identify three components of that pattern, namely linkages within union republics, among union republics, and between the Soviet Union and the outside world. The first two components accounted for most of the post-Soviet decline. Intrarepublican traffic fell in all the new states. This was accompanied by a thinning of the network of linkages which was particularly noticeable in passenger transport. Minor routes, from city bus routes to northern air routes, closed down, and the frequency of service declined elsewhere.

Trade among the former Soviet republics fell by half within two years (Table 13.4). The traffic that remained tended to focus on major border crossings, now equipped with customs posts and all the other paraphernalia of international trade. Minor crossings might simply be closed and their local movements disappear, though many road crossings remained open and virtually unattended for lack of money to pay staff.

Least affected by decline were movements between the former Soviet Union and the outside world (Table 13.4). Many commodity exports

from Russia stabilised or grew again after 1992: total exports expressed in dollars grew by 13% by 1994. Furthermore the network of external linkages began to expand. Ports and airports previously restricted to domestic traffic began to handle international movements, new airline and shipping routes appeared, and cross-border motor traffic grew most of all.

Two other trends in traffic were common to most of the ex-Soviet states. Freight was offered for carriage much less regularly, and in lots of much smaller average size, than hitherto. Both trends reflected pervasive inter-factory indebtedness and the uncertainty of markets, and they exacerbated the problems of an industry used to handling large, predictable shipments.

Trends in traffic varied considerably by mode (Tables 13.3 and 13.5). Russian air transport carried fewer passengers and less freight each year, for total reductions of 63 and 72% from 1990 to 1994. Most of the decline was in domestic traffic. Passenger movements on international routes fell 20% by 1992 but recovered thereafter. Other former Soviet states seem generally to have fared worse than Russia. Ukrainian Airlines cut its daily flights from 1000 a day when it was a

Table 13.4 *Selected Russian exports, 1990–92*

	1990	1992	% Change
Exports to ex-Soviet countries			
Coal ('000 tonnes)	29 153	16 373	−43.8
Fuel oil ('000 tonnes)	22 160	8 801	−60.3
Sawn lumber ('000 m³)	8 772	4 418	−49.6
Cement ('000 tonnes)	4 666	1 974	−57.7
Pipes ('000 tonnes)	2 147	832	−61.2
Trucks ('000)	202	108	−46.5
Tyres ('000)	14 693	4 461	−69.6
Tractors	81 620	33 854	−58.5
Exports outside the ex-USSR			
Coal (million tonnes)	20	18	−10.0
Petroleum (million tonnes)	99	66	−33.1
Natural gas (million m³)	96	88	−8.4
Round timber ('000 m³)	15 000	10 682	−28.8
Wood pulp ('000 tonnes)	389	399	+2.6
Pig iron ('000 tonnes)	2 549	1 931	−24.2
Potash ('000 tonnes)	1 969	4 123	+9.4
Tractors	22 000	9 364	−57.4

Sources: *Narodnoye khozyaystvo RSFSR v 1990 godu. Statisticheskiy yezhegodnik*. Moscow: Respublikanskiy informatsionno-izdatel'skiy tsentr, 1991, pp. 48, 49, 59; *Rossiyskaya federatsiya v 1992 godu. Statisticheskiy yezhegodnik*. Moscow: Goskomstat Rossii, 1992, pp. 46, 47, 57.

Table 13.5 *Passengers carried, common carrier transport, Russia, 1990–94 (millions)*

	1990	1991	1992	1993	1994
Railway	3 143	2 677	2 372	2 324	2 062
Motorbus	28 626	27 302	24 874	24 124	23 438
Taxi	557	526	266	139	98
Tram	6 000	7 619	8 071	8 125	7 644
Trolleybus	6 020	8 005	8 619	9 102	8 751
Metro	3 695	3 229	3 567	4 212	4 224
Maritime transport	16	14	9	6	4
Inland waterways	90	75	44	40	28
Air transport	91	86	63	42	34
Total	48 238	49 533	47 885	48 114	46 283

Source: *Rossiya v tsyfrakh*. Moscow: Goskomstat Rossii, 1995, p. 271.

regional division of Aeroflot, to 100–150 a day in 1993, and air freight in Belarus fell 60% from 1990 to 1992 compared to 44% in Russia.

On the railways, trends in freight and passenger traffic differed sharply. On the Russian railways after 1992 the number of passengers fell much less than the freight tonnage, while passenger-kilometres even rose by 7.5% in 1993 – reflecting tariff changes which favoured the railways over both bus and air transport. Freight movements in Belarus fell 22% from 1990 to 1992, about the same as in Russia, but passenger movements rose 19%. The picture was similar in Kazakhstan.

Exact figures for motor transport are not easy to obtain, since traffic in many places was shifting to the private sector and the collection of statistics was slow to adjust. By 1994 most freight on Russian roads was being carried by private organisations other than privatised state common carriers. Total tonnage fell by nearly a fifth from 1990 to 1992 and more rapidly thereafter. Figures for other ex-Soviet countries seem to be comparable. Passenger traffic also declined heavily. In the case of city transport this seems to have reflected in part falling demand and in part a failure to meet demand, whereas the decline in freight traffic reflected primarily a fall in demand. Motor traffic everywhere is, of course, overwhelmingly domestic. The overall figures therefore hide one area of growth, namely international traffic. In Russia in the first half of 1993, for example, while overall traffic fell by over 40%, international traffic rose threefold, mainly because imports rose more than ninefold. But international traffic accounts for much less than 1% of the total.

Maritime transport, at least in Russia, shows a similar divergence of trends between domestic and international traffic, though the picture is complicated by concomitant shifts from Russian to foreign carriers for much import–export freight and from Russian to foreign employment for much Russian shipping (Tables 13.3 and 13.6). Also, only about half of Russian import–export traffic now moves through domestic ports (Table 13.7), the rest going mainly through Finnish, Ukrainian and Baltic state ports. Thus, by 1994, three-quarters of the tonnage carried by Russian carriers were for foreign charterers, some not involving Russian freight at all. At the same time, foreign maritime shipping companies were carrying most of the Russian foreign trade moving through foreign ports and some of that moving through Russian ports, and they were even participating in Russian cabotage (domestic) activity.

The decline in cabotage especially affected routes serving the north: the 4.9 million tonnes carried in 1994 represented a 39% drop from 1993. On the other hand some northern ports, notably Murmansk, Arkhangel'sk and Kandalaksha on the White Sea, showed exceptional growth in international traffic. The fate of ports in general varied widely. Several Russian Baltic, Black Sea and Pacific ports increased their international traffic, while many minor Russian ports, and major ports in the Baltic states, Ukraine and Caucasian republics, lost traffic.

The most catastrophic declines in traffic in Russia were on the inland waterways (Table 13.3). Virtually all types of traffic were affected.

Table 13.6 *Freight dispatched internally, common carrier, Russia, 1990–94 (million tonnes)*

	1990	1991	1992	1993	1994
Maritime transport	82.80	76.20	70.90	69.10	60.00
including					
Export		20.10	19.10	16.00[a]	
Import		12.40	3.30		
Between foreign ports		0.60	0.30		
For foreign charterers		43.10	48.50	53.10	51.50
Inland waterways	17.50	17.00	16.40	20.10	22.50
including					
Export		9.40	8.00	17.20[a]	18.90[a]
Import		2.40	2.60		
Between foreign ports		4.80	5.70		3.50
Transit		0.40	0.10		
Motor transport	1.63	1.45	1.65	1.68	
including					
Export	0.88	0.76	0.77		
Import	0.44	0.44	0.50		
Transit	0.31	0.25	0.38		

[a] Export and import together.

Sources: *Rossiyskaya federatsiya v 1992 godu Statisticheskiy vezhegodnik*. Moscow: Goskomstat Rossii, 1993, p. 542; *Sotsial'no-ekonomicheskoye polozheniye Rossii*. Moscow: Goskomstat Rossii, 1994, p. 67; Gol'ts, G. and Filina, V., "Strukturnyye sdvigi na transporte", *Avtomobil'nyy transport*, 1995, No. 5: 22–4; Bagrov, L., "Rechnoy transport: itogi, zadachi", *Rechnoy transport*, 1994, No. 2: 3; Bagrov, L., "Rechnoy transport Rossii: itogi i prognozy", *Rechnoy transport*, 1995, No. 2: 17; Tsakh, N., "Morskoy transport Rossii segodnya i zavtra", *Morskoy flot*, 1994, Nos 5–6: 1; "Gruzovyye perevozki v usloviyakh rynochnoy ekonomiki", *Avtomobil'nyy transport*, 1994, No. 9: 19; "Ill Mezhdunarodnaya Vystavka-Konferentsiya ÀSMAP-95", *Avtomobil'nyy transport*, 1995, No. 8: 3; "Itogi raboty morskogo transporta i zadachi na 1995 g.", *Morskoy flot*, 1995, Nos 3–4: 3.

Table 13.7 *Seaports, freight handled, Russia, 1992–94 (million tonnes)*

	1992	1993	% Change 1992–93	1994	% Change 1993–94
Total freight	79.0	70.8	−10.4	60.0	−12.0
Export	20.5	29.7	+44.9	39.0	+27.0
Import	22.6	17.0	−24.8	8.0	−56.0
Cabotage	35.9	23.3	−35.1	12.0	−33.0

Note: Figures for 1993 from MF. Figures derived from GKS are similar except for cabotage, which is 17.9 million tonnes.

Sources: *Sotsial'no-ekonomicheskoye polozheniye Rossii*. Moscow: Goskomstat Rossii, 1994, p. 67; Tsakh, N., "Morskoy transport Rossii segodnya i zavtra", *Morskoy flot*, 1994, Nos 5–6: 3; "Itogi raboty" morskogo transporta i zadachi na 1995 g.", *Morskoy flot*, 1995, Nos 3–4: 4.

Sand, gravel and wood movements, the mainstay of waterway transport, collapsed with the construction industry, and oil traffic declined as production fell and prices rose. Shipments both to and from the north declined as they did in maritime transport. Passenger movements more than halved between 1990 and 1992, and they fell another third by 1994. Only international traffic rose absolutely and relatively, from 3 to over 9% of tonnes carried on the waterways.

In sum, available figures for the post-Soviet period paint a picture of generally falling traffic, except for traffic with the outside world. The Soviet system of linkages was beginning to break up into separate national systems, and those systems in turn were beginning to weaken internally. But an examination of the various transport modes separately reveals more complexity. To understand the evolving situation, we need to examine a number of influential factors.

Influences on traffic and transport

Influences on traffic and transport after the breakup of the USSR can be divided into two overlapping groups, namely those affecting the traffic offered to carriers and those acting directly on the organisation, operation and financing of transport. Both groups can be further divided into influences stemming from international relations, both among the former Soviet states and between them and the rest of the world, and those stemming from internal conditions in the new states. We shall review these influences briefly, then consider some of them in more detail.

Influences on traffic offered

The Soviet economy was tightly integrated, in the sense that most factories received inputs from assigned suppliers and delivered their output to assigned customers. For many commodities there were only one or two production chains for the whole country, and the components might be in any of the republics. There was also little slack, in the sense that there were few goods looking for customers but, if production targets were not met, many consumers seeking unavailable goods. In other words, any disruption of linkages was likely to have a big cumulative impact, and that happened with the breakup of the union. Most of the new states were prepared in principle, and because of a lack of investment capital, to maintain production-chain linkages that had become international, even though they were perceived as sources of vulnerability to political pressure. Willingness in principle, however, was undercut in practice by three major factors.

First, most of the new countries pursued policies to strengthen national control of economic development. These included bans on the export of scarce goods priced higher in neighbouring countries, monetary policies designed to isolate the new national currencies from unwanted outside influences, and, to the extent that capital was available, investment to avoid dependence on foreign suppliers and raise the level of national self-sufficiency.

Second, most of the countries placed a higher priority on earning hard currency than on maintaining links with their ex-Soviet partners. Russia, for example, reduced oil shipments to its former partners. This was in order to restore exports outside the former Soviet Union, following a drop which removed the country's main source of hard currency, and maintain domestic deliveries despite falling production. Deliveries to former Soviet states, which were still paying well below world prices, therefore had low priority.

Third, trade was hampered by economic conditions in the new countries. Even when suppliers of inputs were willing to continue shipments, their customers often could not pay. Some could not offer an acceptable currency, others were not being paid for their own products or could not produce enough of them. Inter-factory indebtedness reached crippling proportions both nationally, especially in Russia and Ukraine, and internationally. Industrial output was reduced by war and civil disturbances, especially in the Caucasus and Central Asia. And while Russia reduced the state's economic role faster than a replacement system could be created, others like Ukraine long continued to subsidise inefficient enterprises and were reduced to printing money, with equally disastrous consequences.

All of these factors tended to reduce the freight offered for carriage among the ex-Soviet states, while increasing the amount offered for carriage outside the system as long as production sufficed. Passenger flows were also affected by the new political geography. Business and government travel between the ex-Soviet Union and the outside world inevitably grew. Tourist traffic among the ex-Soviet countries, by contrast, fell sharply. Many Ukrainians working in the Russian north returned home for good, reducing the toing and froing of earlier years. And transport had to cope with refugee flows of considerable volume but irregular occurrence.

Changing internal conditions in the new states also affected their domestic traffic. It fell rapidly in consequence of collapsing industrial production and a drastic decline in consumer purchasing power.

Direct impacts of the new political geography on transport

When the USSR broke up, its transport systems were divided among the new states. Decisions had to be made on the allocation of mobile assets, from ships to railway rolling stock. Transit arrangements had to be made for Russian goods moving by rail or pipeline to Baltic and Ukrainian ports, and for Central Asian, Armenian and Azerbaijan freight to reach ocean ports at all. The continuation of former union services, such as the Trans-Siberian Container Landbridge from the Far East to Western Europe and the international services of Aeroflot, required new agreements. So did the continuation of training and servicing arrangements. For example, servicing facilities for most types of aircraft were concentrated at one or two factories, and only two establishments trained helicopter pilots. If agreements could not be reached, the new states had to fend for themselves. In the case of transit, of course, the interior states could be virtually held to ransom by their partners.

In addition to these immediate concerns, co-operation among the new states could be hampered by divergent national transport policies. Transport authorities which remained as branches of government might find themselves dealing with privatised concerns in other countries, and the two might differ completely in access to financing, the nature of government control and requirements, and preferred standards and types of equipment.

Direct impacts of internal conditions on transport

Most of the new states differ in transport organisation and the role of government control. For example, Kazakhstan has one ministry of transport, but Russia has two. Privatisation has gone much further, and regions have been able to secure more independence of action in Russia than in most of the other states.

Differences in organisation and control tend to be associated with differences in financing and government tariff policies. Private companies may be able to switch assets from domestic to foreign trade in order to increase profits, while government agencies may be forced to cross-subsidise in order to preserve money-losing services. Russia has differentiated among modes in its tariff regulations, causing traffic to switch from one to another. States varying in their attitudes to private enterprise may also vary in their ability to attract foreign investment in transport, though variations in general economic conditions, such as currency stability, have probably been more important.

Almost everywhere, fuel is scarce and much more expensive than it used to be. Also, even more than under *perestroika*, the criteria for satisfactory performance in transport have changed for both people and machinery. Such considerations have produced a widespread dissatisfaction with existing transport equipment, and indeed the removal of some from service on the grounds that it has no prospect of generating a profit. The various states and their transport organisations vary greatly in their ability to buy replacements. The states also vary in the sophistication of their banking systems and hence in their ability to arrange financing.

The states differ in their levels of law and order as well. War has disrupted transport operation and maintenance in the Caucasus and Central Asia. Criminal activity is affecting port operation in the Russian Far East. Weak central governments, most evidently in Russia, are unable to enforce safety standards.

In sum, the breakup of the Soviet Union has had widespread ramifications in transport. What was observed for traffic patterns is to some extent true also of transport. Because of both inter-state relations and differing internal conditions, the present trend is divergent. In other words the various parts of what was a relatively homogeneous transport system – or set of systems, since the various modes were not well integrated in Soviet times – are changing into separate national systems. They share a number of trends and they are bound to retain close ties, but it would be difficult for them to co-operate as closely as they used to, now that their parent states have different economic and political systems and priorities and are experiencing different trends in economic development.

Some of the influences on transport will now be examined in more detail.

Transport organisation

The former Soviet countries, as stated above, vary in their approach to transport organisation. Russia has one transport ministry for the railways, which remain under government control through the ministry almost in the Soviet fashion, and one for the other modes. The latter and the single ministries in the other new states are more similar in concept to transport ministries in Western countries. That is to say they are concerned with regulation and the implementation of government policy, while transport operations are run by separate entities, which might be government corporations or privatised companies. The term "privatisation" covers situations ranging from complete separation from government to partial share distributions with the government retaining effective control. Railways in general have remained in government hands, and much motor transport has been privatised, while policy towards air transport has varied. Russia has privatised regional divisions of Aeroflot, whereas smaller countries like Azerbaijan, Turkmenistan and Uzbekistan have kept former Aeroflot assets as one state company. In most countries air, motor and water transport have also seen the emergence of private companies with no previous connection to government. Change has been most complex in Russia, and the following discussion will focus mainly on that country.

Russia has pursued privatisation in transport more assiduously than other ex-Soviet countries, though not as assiduously as it has in other branches of the economy. Even leaving aside the railways there has been a reluctance to privatise transport infrastructure, and a number of transport operations have been classified as special cases needing continued government control. Thus in motor transport over 60% of some 3600 enterprises remained under the control of one or other level of government in mid-1994, and they were still required to assign trucks to help bring in the harvest – 43 000 of them in 1993, including 9000 sent out of their home regions. In sea transport the government has cancelled or delayed several port privatisations, especially in the Far East. And both maritime and river shipping companies handling Arctic and Far Eastern traffic are considered to be special cases, in which the government is retaining a controlling interest for the present. This is presumably in order to ensure deliveries to the north at a time when the government lacks the money to subsidise them directly.

The favoured form of privatisation has been the joint-stock company. Of two permitted forms, the most frequent choice has been the "open" form, in which 51% of the shares are distributed by closed subscription to employees and managers. Of the remainder, 20.0–25.5% remain with the state and the rest can be bought on the open market. Clearly this approach enables current managers and employees to retain control, and to some extent it preserves the existing scale of organisation, though the government has insisted on, for example, the separation of seaports and major airports from the shipping companies which formerly ran them. Many transport enterprises have opposed or tried to circumvent such efforts to promote competition. This can be interpreted as an attempt to preserve monopoly power, but there are more laudable reasons for their stance. In the present turbulent economy a diversified organisation has a better chance of finding some profitable activity which will keep it going until restructuring is complete. Naturally those transport organisations which depend on ancillary activities to subsidise transport operations are reluctant to give them up.

Some companies have indeed managed to avoid splitting up. More small firms have been created in motor transport than in any other mode, but the regional common-carrier organisations in Primorskiy *kray*, Bashkiria and a number of provinces managed to privatise as wholes. With 30 subdivisions and 9000 employees, and a manager trained in the Columbia University business school, the Primorskiy *kray* motor transport company is a formidable opponent for any newcomer. Even when transport organisations have been split up, that does not necessarily mean that all ties have been severed. Many companies have bought stock in former subsidiaries. In maritime transport, shipping companies have interests in an average of 15 ports and smaller businesses.

Air transport has posed particular problems because of its tightly regulated international environment. In the USSR one company, Aeroflot, ran everything from international passenger routes to airports and crop dusting. Aeroflot was broken into over 200 companies, including airports, former regional divisions of the original company, and Aeroflot International Airlines, set up in Russia to handle international flights and administer the foreign landing privileges held by the old Aeroflot. Over 100 offshoots with international licences, in Russia and the other ex-Soviet states, at first flew abroad under the Aeroflot flag. But some incurred hard-currency debts which Aeroflot was forced to pay, and the arrangement with non-Russian airlines was terminated. Even among Russian companies, Aeroflot's domination of flights abroad has fallen to below two-thirds, because of the rise of private airlines with no previous Aeroflot connections. By 1995 no less than 400 Russian companies had the right to fly internationally. Aeroflot itself remains essentially a government organisation: 51% of the shares are held by employees and 49% by the government.

Air transport can also illustrate the problematic nature of the Soviet legacy. In competing with the new private airlines, ex-Aeroflot divisions start with the disadvantages of overmanning, a lack of essential skills, obsolescent equipment and a network of unprofitable routes. Ukrainian Airlines, for example, inherited only short- and medium-haul aircraft, and crews untrained for international work. International flights from Kiev had been handled by crews and aircraft sent from Moscow. Despite continuing subsidy from the Ukrainian government, the company had to cut the number of daily flights by nearly 90%. New private airlines have captured much international business, the only growing sector. In Russia they have also been capturing the most lucrative domestic traffic, namely business travellers, by using Western aircraft, offering superior service and maintaining high safety standards. They have been able to achieve high load factors while charging much

more than the former Aeroflot divisions, which generally have poor reputations for safety and service. As might be expected, few private airlines have emerged in the poorer countries of the Caucasus and Central Asia.

Transport organisation is still in transition in the former USSR and especially in Russia. Many of the multitude of companies will not survive. And the demise of some will result not from competition, but from the attitudes of unsophisticated shareholders. Employee-shareholders have tended to demand higher wages rather than capital investment, assuming that the government will continue to provide new equipment. Other shareholders have refused to permit stock issues as a way to raise capital for expansion, because they fear losing power in their company.

Transport tariffs and financing

Transport financing during the 1990s has been characterised first by reductions in, and often the disappearance of, government subsidies; second, especially in Russia, more reliance on tariffs and ancillary activities to finance operations and on foreign investment to finance development; and third, shifts in the responsibility for setting tariffs, away from central government. Also, patterns of taxation have changed. In addition to taxes levied by the central government, regional and local authorities have the right to levy their own taxes. Levels vary, and there is no formal consultation on the cumulative effects of the several layers. All of these changes have resulted in differential changes in tariffs among the various modes and in turn, therefore, shifts of some traffic from one mode to another; the rapid decline of some types of traffic, because unsubsidised tariffs exceed customers' abilities to pay carriage on any mode; differential abilities to finance investment, among the various modes and within them; and considerable interregional differences in tariffs – for example, in those for suburban transport, which are not centrally controlled – as companies try to adjust to local levels of taxation. In other words the shift of financing patterns has brought immediate changes in the geography of traffic flows and will probably bring longer-term differentials in the fortunes of different modes and routes.

Freight tariff liberalisation began in Russia in 1991 and 1992. The government began to free fuel prices in 1992, and the increased costs to transport led to tariffs being freed on air, motor and river transport, with the restraint of a 35% maximum mark-up over costs. Railway and domestic maritime tariffs remained under central control. An attempt was made in December 1991 to free some railway tariffs too: individual railways were allowed to set tariffs for local movements. But some railways abused their monopoly position, and the right was removed within a year. Maximum rates are now set centrally for Russia, though railways can charge less. Tariffs among ex-Soviet countries are set by the Council of Administrative Heads of Railway Transport, and in mid-1993 they were averaging 63% higher than Russian domestic tariffs – obviously a disincentive to the continuation of intra-CIS trade. In Russian maritime transport, tariffs for domestic movements, port handling charges, and ice-breaking charges are all centrally controlled, though coefficients applied to the basic scale vary by port and shipping company.

Passenger tariffs remain centrally regulated on the Russian railways (except for suburban services), on interprovincial river routes and on domestic maritime routes. Air tariffs were freed at the beginning of 1993 on scheduled internal air routes, with the constraint of a 20% mark-up over costs. Local authorities were permitted to set tariffs for intraprovincial passenger movements by all modes, including city and surburban services, and on interprovincial bus routes.

The mixture of control and freedom soon changed the relative level of tariffs. In passenger transport it brought a shift to the railways. In 1991, over a distance of 1500 km, air fares were 77% above rail fares. By May 1993 the figure was 443%, though the differential has since fallen again. Bus fares moved from 32 to 61% above rail fares in the same period. Some fares were raised to prohibitive levels: the demand for air tickets fell by over 40% early in 1993, and a survey at a Moscow airport showed that holiday travellers and those visiting relatives had declined to 6% of the total from 40% in 1988. The transport authorities were not necessarily attempting to

gouge. In most cases even the new charges are too low to cover costs, and Russian domestic passenger services as a whole are subsidised from freight or from international services. Railway passenger services, excluding suburban, covered only 50% of their costs in 1993, and bus services 10–15%. Similar situations prevailed in Kazakhstan and Belarus.

Even if tariffs are set to cover costs, transport companies still lose money because customers cannot pay their bills, often because of the cessation of government subsidies. In Russia the loss of subsidies has had the most traumatic effect on northern transport. The government has promised to continue subsidising the northern economy but so far has lacked the money to do so effectively. The transport companies responsible for delivering supplies were not paid in full in 1992 (the government itself being among the offenders, and the Defence Ministry one of the worst), so in 1993 they demanded payment in advance, until ordered to work without it by President Yeltsin. But the suppliers had not been paid either, so they sent less freight than usual, and the shipping companies lost more money through having to dispatch their ships half-empty. In 1994 the government set up a federal fund to meet the costs. Lack of money has affected investment as well as operations. The railway to Yakutsk, long under construction, should be finished as far as Tommot in 1996, but there is no prospect of completing the final stage. Loss of subsidy has forced northern airports, maritime, river and air routes, and even some airlines to close down, except where mining enterprises or communities have been able to take them over. By 1995, when many Russian airline companies were finally approaching financial stability, the remaining loss-making companies were largely concentrated in the north. The same was true of maritime transport.

Transport subsidies have not entirely disappeared in Russia. Even companies which can cover operating costs lack the resources to buy new equipment. In 1993 not one Russian airline which relied on internal routes was able to afford a new aircraft without government assistance, and matters were scarcely better in 1994. In both years the state helped several airlines buy new Russian aircraft. Even companies with foreign business needed assistance: Aeroflot received help to lease a Boeing-767 in 1994. But in 1995 funds virtually dried up. The government hopes to continue helping the airlines, but more to lease modern foreign aircraft for a few years, until new Russian ones are ready and the airlines can afford them, than to buy. Other modes have also received government attention. A series of investment programmes has been announced, beginning in 1992 and ranging from "Russian Internal Waterways" and a road-building programme to "The Revival of the Russian Commercial Fleet", which envisages the purchase of several hundred ships by the year 2000 through a combination of federal subsidies, foreign financing and other sources. But there is so little government money that the programmes have become largely instruments for generating and directing private domestic and foreign investment.

Subsidies to transport in other ex-Soviet countries vary a great deal, depending on government resources. Probably the best-supported airline is that of Turkmenistan, since the government has good revenues from natural gas exports, and the worst-supported is that of Tajikistan. Leasing aircraft is popular in the other ex-Soviet countries too. Ukrainian Airlines and the Azerbaijani State Airlines have both been leasing Boeings.

Foreign financing has been forthcoming most readily for airports and seaports dealing with international traffic, though less so for the ships carrying the traffic. International banks have justified their stance on the grounds that commodities from the ex-USSR need to reach world markets, but there is already enough shipping in the world to carry them. Some financing for ships has been obtained abroad by companies engaged in international traffic. In Russia, the Sakhalin Company has turned itself from a regional to a major international carrier by using bank credits from Britain, France and Germany. One effect of foreign financing is that many Russian companies have been foreign-flagging their ships – reregistering them under companies set up abroad for the purpose – in order to use them as collateral for loans. Foreign banks are nervous of accepting Russian-registered ships as collateral. Financing for ships to operate within Russia is lacking, so the number of passenger vessels is

declining and services have been cancelled. Other ex-Soviet countries have had to exercise ingenuity to obtain funds. Georgia, for example, has set up joint companies with British and Greek investors to refurbish and operate inherited Soviet vessels.

Financing problems have compelled transport companies to focus on two sources of revenue: ancillary services and foreign business. Ancillary services may be connected to the transport business directly (e.g. superior-service passenger trains), indirectly (e.g. hotels), or not at all (e.g. the use of maintenance workshops to manufacture garden furniture). In 1992 the Russian railways obtained almost half their profits from them. International business is no less attractive, because dollars can effectively offset high domestic costs. In 1992 the North-West River Shipping Company obtained 80% of its profits from foreign traffic, which accounted for under 10% of tonnes moved. The incentive to shift resources out of domestic transport into ancillary activities or international business is common to air, maritime and river transport.

The division of assets and external economic relations

Considering the economic situation in the new states after the dissolution of the USSR, it would have seemed sensible to continue joint operation of transport, at least for a few years. That, however, was made difficult by nationalist sentiments and mutual mistrust. Though recognising the need to co-operate, all the new states set about creating their own independent transport systems. Especially strong was the desire for access to international markets which could not be impeded by former fellow-republics of the USSR. The western peripheral republics soon began to see high transit charges as both a source of income and a weapon against Russian efforts to raise the price of the raw materials they needed. Lithuania was accused by Russia of raising transit charges to Kaliningrad in order to divert traffic to its own port of Klaipeda. Turkmenistan's access to foreign natural gas markets depended entirely on the good graces of Russia. Concerns of this nature led to initiatives relating to ports, merchant shipping, air transport, highways and pipelines.

Soviet policy favoured specialised ports in both the Baltic and Black Sea basins. Thus oil exports overseas went mainly through Ventspils and Novorossiysk. Dissolution soon gave rise to plans for independent facilities. Russia announced that it would build a superport near the Estonian border and expand Vyborg and St Petersburg instead of continuing to use ports in the Baltic states, and Ukraine invited Dutch and British firms to tender for building an oil terminal at Odessa to enable the import of up to 40 million tonnes a year by sea rather than by pipeline from Russia.

The new state most concerned about ports has been Russia, since it relied heavily on Baltic-state, Ukrainian and Georgian facilities. Russia has about 40 ports with a total capacity of 166 or 187 million tonnes according to different sources. That amounts to somewhat over 40% of total USSR capacity prior to breakup, and 54–64% of the capacity needed by Russia in recent years. Eleven of the ports are equipped for foreign trade and can handle about half the country's current needs. In addition Russia has lost a much larger proportion of its capacity to handle certain commodities, including potash, oil, compressed natural gas, sugar and grain. Furthermore the ports lost were among the most modern in the USSR, the peripheral republics having the best locations relative to most trading partners. Of Russian ports, 60% are too shallow for modern ocean-going ships, and most are poorly equipped.

The new Russian Baltic facilities, the ultimate capacity of which has varied in different statements from about 40 to 160 million tonnes, will be able to handle almost all types of traffic, including train and truck ferries, and will virtually eliminate the need to use Baltic-state ports. In the meantime cargo is being diverted to Murmansk, Kandalaksha and Arkhangel'sk on the White Sea – a measure of nationalistic determination, considering the extra costs involved. The approach to port development in the south has been much the same. Russia retained only 4 of 17 ports on the Black and Azov seas, leaving a deficit in handling capacity of over 40 million tonnes. Novorossiysk and Tuapse, the only Black Sea ports equipped for foreign trade, will add 35

million tonnes of capacity between them. Other ports could be created or expanded on the Sea of Azov and the Taman' Peninsula, either for import–export traffic or to relieve Novorossiysk and Tuapse of domestic traffic.

In addition to seaport expansion and construction, the use of river ports to supplement Russian foreign trade capacity has been pursued by the river transport companies. In 1990 they gained over a third of their profits from river–sea vessels – those able to operate both on the waterways and in coastal waters – which accounted for only about one-fifteenth of the tonnes carried. It has been claimed that river ports could easily add over 15 million tonnes to Russia's export capacity, including 9 million tonnes from upstream ports. In addition to the Baltic and Black Sea basins, they could supplement Russia's only remaining Caspian seaport, the poorly equipped Makhachkala, and are already handling substantial quantities of trade with China, Korea and Japan from the River Amur. Like the seaports not previously used for foreign trade, they need new port equipment, customs facilities, border-guard accommodations and, usually, better rail access.

The corollary of Russia's initiatives is reduced traffic for Baltic-state and Ukrainian ports. The impact is enhanced by the evident preference of Belarus for Polish ports, and in the case of Ukraine, perhaps temporarily, by United Nations sanctions against Serbia, which have reduced Danube traffic to less than half its former level.

Other countries too have been trying to safeguard their future. Kazakhstan and the Central Asian countries have signed port-use agreements with Iran, in addition to investing in Far Eastern ports to try to ensure Pacific access. Armenia has a similar agreement with Turkey for use of its Black Sea ports. And Turkmenistan has been trying to establish a secure pipeline route westwards to a port from which it can export its natural gas.

The allocation of merchant ships to the new states was by port of registry, which was usually a ship's normal base of operations. This left Russia with a fleet able to carry only 30% of its foreign trade cargoes. It became short of grain carriers, lighter-carriers, passenger vessels, refrigerated vessels and small- and medium-capacity tankers, the last two going to Latvia in larger numbers than that country could use. Some were designed for Arctic use and have since been chartered back by Russia. Ukraine also became short of tankers and in 1993 announced plans to build 40 of them.

The new states' interest in improving access to the outside world extends beyond ports and shipping. The Baltic states envisage an improved highway link to the trans-Europe network, while Ukraine hopes that roads from Berlin and Krakow in the north, and Lisbon and Budapest in the south, will converge on Kiev and continue to Central Asia. So far the Baltic states have done most in practice to shift their international links away from Russia. By 1993 they had largely re-oriented their airline routes, whereas most of the other states had increased the proportion linking them to Russia. Other Baltic countries, especially Sweden and Germany, are the favoured partners.

The division of transport assets gave rise to many problems in addition to those of maintaining and expanding foreign links. Construction and repair facilities were scattered among the various Soviet republics with no regard for local needs. Thus Russia was left with inadequate capacity to build and repair maritime vessels, railway locomotives and freight cars. Its principal shipyards, on the Baltic, were specialised in a few types of vessel no longer in demand, such as nuclear ice-breakers, floating canneries, research vessels and naval ships. In 1993 it had a deficit freight-car repair capacity of 13 400, while Ukraine had a surplus of 5700, Kazakhstan a deficit of 2700 and Armenia no repair facilities at all. Similar situations prevailed in other branches of transport construction. Soviet civil aviation used 28 types of aircraft and 30 types of engine, and servicing facilities were both specialised and scattered among the republics. Thus the Il-62 airliner, used on long-haul routes in Russia and internationally, was serviced in Tashkent, while the Yak-40 and Tu-134 medium-haul aeroplanes were serviced only in Belarus.

Training facilities for transport personnel were also both specialised and widely scattered. Most training facilities for railway administrators were in Russia, but most railway engineers were trained in Belarus. Helicopter pilots were trained in Ukraine, except for a few in Omsk.

Duplication of facilities seems foolish when all the states are in financial difficulties. The states with surplus capacities for manufacturing, servicing or training would like to use them to earn money, so long as they are sure of being paid. On the other hand their potential customers worry about being vulnerable to economic pressure and about the scarce dollars they might have to pay. And peripheral states are understandably nervous when some Russians advocate close co-operation provided that everyone accepts the "natural dominance" of Russia. In the event, import substitution has generally been favoured if the opportunity has arisen. By 1994 Russia was already producing aircraft, buses, suburban electric trains, railway hopper cars and cement carriers, and spare parts for railway equipment, formerly imported from Ukraine, Hungary, Latvia and elsewhere. Kazakhstan was adapting existing factories to produce railway equipment. In Soviet times it had imported 95% of what it needed from Russia and Ukraine. One continuing problem is that factories which are free to make their own decisions have generally preferred to build for foreign markets if possible, in order to earn hard currency. This has been particularly true of Russian shipyards.

Sometimes co-operation is virtually unavoidable. The Trans-Siberian and South Siberian railways, for example, cross Russian and Kazakh territory indiscriminately. The Trans-Siberian Container Landbridge requires co-operation between Russia, Kazakhstan and Belarus. But even leaving such cases aside, the moves towards more independence have been offset by a number of initiatives to maintain close co-ordination, if not integration, and avoid duplicating assets and work. By mid-1992 there were inter-republic agreements on rail, air, automobile and water transport. In addition, individual organisations formerly of all-union significance have tried to re-establish old ties. They include the training institute for railway engineers in Belarus, the Antonov aircraft company in Ukraine, which is linked to several factories in Russia, and the Zheldorremmash railway equipment repair company, based in Russia but with associated companies in Belarus, Kazakhstan, Central Asia and Latvia.

Efforts to co-operate have encountered difficulties. The travails of Aeroflot International

Airlines, leading to the severing of connections with non-Russian airlines, have already been mentioned. Efforts to co-operate on the railways have also been hampered by inter-state mistrust. Through runs of locomotives and rolling stock, i.e. without any change of equipment at international borders, are both technically feasible and economically logical, but the new states have been reluctant to maintain them, fearing that in a time of severe shortages and poor maintenance any good equipment sent abroad will stay there. Border crossings present other problems too. All need extra sidings, customs posts and associated facilities, so there is a strong incentive to consolidate flows and shut some crossings. The same is true of highways. The creation of borders therefore has a cumulative separating effect on the national transport systems.

Equipment, supplies and capital investment

Equipment problems of the successor states do not stem entirely from the ways in which assets have been divided up. No less serious are the age of much transport equipment and its unsuitability for current needs. The average age of both maritime and river vessels is now approaching 20 years, and over half of Russia's aircraft should be scrapped by the year 2000. Figures for railway and motor transport equipment are only slightly better.

"Unsuitability for current needs" can mean several things. Much ex-Soviet air, sea and motor transport equipment is unacceptable abroad. The current mainstays of Aeroflot's fleet are too noisy and polluting, and many ships too no longer qualify for entry into foreign ports with strict pollution standards. And even new heavy trucks from ex-Soviet manufacturers fall below current environmental standards for operation in the European Union.

The operating costs of Soviet-era equipment have become a major concern, as transport organisations have had to both compete and strive to keep their prices below levels which eliminate traffic altogether. Labour and fuel are particular problems. The overmanning common in the Soviet economy was especially prevalent in water transport. Both maritime and river ships carried much larger crews than their Western counterparts, partly because unreliable machinery needed constant maintenance. Now crews are smaller, and many Russian crews are working abroad. Fuel costs are a far larger share of total costs than before on domestic routes, and vessels which consume heavily cannot operate cheaply enough to retain traffic. They are also a liability when fuel is hard to obtain at any price. On the Russian rivers over 400 passenger vessels had been laid up by 1993. They were mainly hydrofoils, air-cushion vessels and large cruise ships. Among freighters, tug-and-barge systems had replaced motorships at fuel savings of up to 30%, though the largest tugs have been laid up too.

High fuel costs and poor availability are affecting all forms of transport, and the unpredictable relative movement of costs makes it difficult to plan for the long term. For example, up to 1994 railways were switching traffic to electrified lines and using scarce financial resources to electrify more of them, especially the two remaining non-electrified sections of the trans-Siberian east of Lake Baykal. At 1993 prices in the Russian Far East, electricity was one-sixth the cost of diesel fuel. Then electricity prices rose relative to diesel fuel prices, casting doubt on the policy. Airlines have been switching to leased Boeings or Airbuses rather than continuing to use thirsty Russian aircraft, and many airline flights have been cancelled for lack of fuel or because passengers cannot pay fares high enough to pay for it.

Perhaps the greatest unsuitability problem is that of the ships built for Arctic work. Soviet Arctic transport was heavily subsidised, and it was required to move millions of tonnes of freight without fail in harsh conditions. Ships, from nuclear ice-breakers to diesel-electric freighters, were therefore built for strength, durability and versatility. Arctic ice-breaking freighters, for example, commonly incorporated several technologies for loading and unloading, including crane helicopters and air-cushion lighters. Economical operation was not a priority. Under post-Soviet conditions such vessels are often seen as a costly liability by the operating companies, especially the Murmansk Maritime Shipping Company, the operator of the nuclear fleet. An indication of the scale of the problem is that the first ships which were to be built under the Programme for Renewal of the Russian Commercial Fleet were all for Arctic use.

Immobile equipment in most branches of transport is no better suited to current conditions than mobile equipment. Few seaports or airports come up to international standards, and delays are frequent and costly. There are few all-weather airports away from the main routes, and many minor ports handle ships only at road-steads. Russia is considered to have about a third of the minimum length of public roads it needs, and investments in construction and maintenance have been declining year by year.

The obvious answer to equipment problems is capital investment, but, as we saw in the section on finances, few transport companies can afford to buy expensive items like aircraft and ships. Those that can, with or without government help, tend to avoid ex-Soviet products, which often embody precisely those qualities which the operators now want to avoid. But the successor states do not want their transport equipment industries to die. That is why the Russian government is helping airlines lease rather than purchase foreign aircraft, hoping that the next generation of Russian machines will meet international standards. Several international joint ventures are devoted to ensuring that they, and other new transport equipment, will do so. Russian aircraft are being fitted with British and American engines and American avionics. Buses are appearing with Russian bodies and Swedish chassis. And the latest cars owe much to development work by the Porsche company.

The urgency of need for new equipment varies among modes. Traffic on river transport has fallen so much that many companies have spare vessels. The main demand is for river–sea ships to work abroad, and many are being converted to meet international requirements. Declining traffic has also helped the railways. Capacity has fallen because train weights have had to be reduced to what can be carried by worn-out or poorly maintained track, and because frontier delays, poor industrial discipline and mechanical failures have reduced the stock of freight cars. In air transport the most urgent need is for new equipment to serve international routes. On internal routes the existing equipment could serve for some years. In road transport, rising tariffs and the collapse of manufacturing industry and construction have kept down the need for new lorries, with the exceptions of big trucks for international use and small local delivery vehicles, neither of which were made in quantity in the USSR. Passenger vehicles, however, are a serious problem. Equipment is wearing out and not being replaced, and the demand for service has not declined as it has for freight services. By 1994 Russia had 55% of the buses it needed, and 30% of the remaining ones should have been scrapped. Over 3000 bus routes have been deleted, and the intervals between buses on the remaining routes have increased greatly. The situation was expected to improve as new Russian production facilities came into operation, replacing the imports from Hungary and Ukraine on which the country previously relied. At the end of 1995, however, buses were available but operators could not afford to buy them.

Conclusions and forecasts

Changes in transport since the breakup of the Soviet Union have been complex, but some generalisations can be made. First, while traffic within and among the former Soviet republics has generally declined, traffic between them and the outside world has assumed relative prominence. International transport was a minor component of air, rail and sea transport in Soviet times. Since then, assets have been switched from domestic to international transport, and there has been more new investment in the latter. Companies engaged in international work have on the whole been much more prosperous than those restricted to domestic movements. The former have therefore been able to maintain or expand their operations and renew their equipment, while many of the latter continually hover on the brink of bankruptcy and have been forced to abandon services. Viewing recent processes in historical context, we could say that tsarist times saw the gradual creation of an Imperial transport network, represented primarily by railways and strongly oriented to serving foreign trade. The Stalinist era brought an internalisation of the economy, and the transport system was modified to meet its needs. After Stalin the level of internalisation was softened, first to improve links to Eastern Europe and then to permit more Soviet participation in world trade. Dissolution of the Soviet state divided one transport system into 15, and their separateness seems to have been growing since then. At the same time they have weakened internally and strengthened their links with the outside world. To some extent recent processes can be seen as a reversal of Soviet internalisation, but the separation into national systems and the internal weakening are new in the transport history of the region.

Second, there has been taking place a shift of traffic among the transport modes. In part the reasons are immediate. Thus the growth in railway passenger traffic reflected a diversion from air and motor transport when deregulated ticket prices rose much faster. But some shifts may reflect long-term secular change. Thus the decline in railway freight traffic has been seen by some analysts as both a reflection of immediate national economic problems and an expression of probable long-term trends in the economy in general and transport in particular. The latter include expected lower material intensity of post-Soviet industry, a shift away from the big, regular single-factory shipments associated with state-monopoly production, and consequent change in the intermodal proportions of transport in favour of trucks. These forecasts assume that Eastern Europe will become more like Western Europe, an assumption that could be questioned on the grounds that severe physical conditions make Siberia at least more suited for continued reliance on railways, and that even in recent times Soviet industry was much more commonly sited for good rail access than in the West. Furthermore, the cost of bringing ex-Soviet road systems up to the standards of Western Europe, in terms of quality and coverage, would be horrendous. Improving multimodal transport might be a better solution.

Third, competition has been introduced into transport quite effectively, at least in Russia. Private companies have had considerable impact in air, motor and sea transport, and even river transport has lost business to former customers who have established their own fleets rather than continue to rely on common carriers. The combination of competition, deregulation of tariffs, and the reduction or disappearance of government subsidies has already eliminated some airline, coastal-shipping and city transport routes altogether, and further reduction in services seems inevitable. Reduction in the number of companies also seems inevitable, with consolidation of airlines most likely in the immediate future. At present the railways, still subject to close government control, are being kept aside from these processes, except that they are acting as a repository for traffic discarded from other modes, indeed a kind of safety-valve as far as long-distance public passenger transport is concerned. At the same time their relative lack of exposure to competition may not bode well for their long-term future, as other modes are forced to become more efficient. Where the railways are not able to compensate for traumatic change in other modes is in city transport. With the disappearance of so many bus routes, a lack of taxis, and severe financial pressures on the metro systems, it will be interesting to see how cities built up in reliance on public transport manage their traffic flows in future. The USSR had few private cars, but their number in Russia has been growing rapidly.

Fourth, the servicing of remote areas, especially in the north, appears to be in steep decline. If present trends continue, it seems likely that Russian northern transport will eventually look more like that of northern Canada and Alaska than it has in the past, with much smaller volumes of traffic than in Soviet times and consequently a greater reliance on air transport. One would not expect complete convergence, given the infrastructural inheritance from Soviet times and the different physical conditions, including the greater usability of Russian northern rivers and Arctic waters.

Fifth, Russia at least seems to have decided that it must manufacture transport equipment sufficient for its needs. This may represent an immediate response to shortage of foreign currency, but there appears to be a longer-term resolve not to rely on foreign suppliers. If that is indeed the case protection of industry is likely to continue, especially in aircraft construction. That in turn could mean pressure on airlines to buy Russian, which further raises the question, to what extent the freedom from government interference, which is supposed to be making Russian companies more competitive in world terms, is likely to remain a reality in the long term. The same question is raised by the programmes for building ports and restoring the merchant fleet, since it has not been stated how shippers are to be persuaded to use Russian ports and ships.

Finally, though not discussed in this chapter, there has been a decline in the transport literature of a particular type of writing characteristic of Soviet times. This purveyed grandiose visions of the long-term evolution of Soviet transport. New forms of transport, such as vertostats (a combination of airship and helicopter) and monorails for the north, and immensely expensive schemes such as east–west railways across northern Siberia and all-year operation of the northern sea route, were prominent in such writing. Now realism has set in, with a few exceptions. One recent manifestation of the old style extols the virtues of the "ecranoplan", or supersurface craft, for year-round use along northern rivers. It resembles an aircraft but is designed to fly just above the surface where less lift is required (Smerdov and Lyubimov, 1994). Another scheme is that for a railway tunnel under the Bering Strait, linked to transcontinental lines at both ends. Though put forward partly by Western interests as a potential joint Russian–American venture, it sits squarely in the Soviet visionary tradition. At the same time it should be recognised that Russian writers are enthusiastic about the scheme in part because they see it as a way to get foreign investors to pay for improved transport in the extreme north-east (Bugromenko, 1994).

The most recent indications of future trends are of a more sober nature. First, even the most optimistic domestic traffic forecasts assume that recovery from the precipitous decline of the past few years will be slow. It may take until 2010 to reach the peak levels of late Soviet times, and some routes may never recover, including air and river routes serving the north and many high-speed river passenger routes. Second, many of the ex-Soviet countries, not least Russia, are obviously having second thoughts about the pace at which Westernisation of their economies is taking place and about their degree of involvement with Western capital. The nationalism which has tended to isolate them from one another is also being modified by economic considerations. If these trends continue, it seems quite likely that economic co-operation within the ex-USSR may revive somewhat in the next few years. Any such changes will obviously have profound effects on transport. They could even reverse the current trend towards further separation of the national transport system, though if that did happen, one would expect it to do so selectively within the group of countries.

References

Bugromenko, V.N. (1994), "Proyekt, dostoynyy velikikh derzhav" (A project worthy of the great powers), *Zheleznodorozhnyy transport*, No. 3, March: 16–18.

North, R.N. (1991), "*Perestroyka* and the Soviet transportation system", in Bradshaw, M.J. (ed.), *The Soviet Union: A New Regional Geography?*, London: Belhaven Press, pp. 143–64.

North, R.N. (1995), "Transport", in Shaw, D.J.B. (ed.), *The Post-Soviet Republics: A Systematic Geography*, London: Longman, pp. 66–87.

Smerdov, V. and Lyubimov, V. (1994), "Tekhnicheskiy progress – na sluzhbu naseleniyu Kraynego Severa" (Technical progress in the service of the population of the Far North), *Rechnoy transport*, No. 3: 26–8.

The factual information in this chapter has been obtained largely from 1990–94 issues of the following Russian journals:

Avtomobil'nyy transport (Motor Transport)
Grazhdanskaya aviatsiya (Civil Aviation)
Morskoy flot (Merchant Marine)
Rechnoy transport (River Transport)
Zheleznodorozhnyy transport (Railway Transport)

Index

Index compiled by Geoffrey Jones